THE GENERAL THEORY OF INFORMATION

Origin of Truth and Hope

A Book of Wisdom for Humanity

We are more concerned with people not hurting each other than with science.

—BACH & BELARDO

THE GENERAL THEORY OF INFORMATION

Origin of Truth and Hope

By

Dr. Christian Bach

and

Dr. Salvatore Belardo

University at Albany, SUNY, USA
1999–2012

in collaboration with

Eva Ristl, M.Sc.

University of London, UK

Copyright © 2012 Dr. Christian Bach

All rights reserved.

ISBN-10: 1470026139

EAN-13: 978-1470026134

Library of Congress Control Number: 2012902977
CreateSpace, North Charleston, SC

DEDICATED TO:

The People

Hope is the fundamental message of this book.

This book is dedicated to solving global hunger and poverty in one lifetime.

Contents

Prologue — x
Preface — xiii
Hope Is the Driver of All Action — xv

Episode One — 1

1. The Whole Story — 3
2. Bell's Speakable and Unspeakable — 7
3. A Singularity without a Place to Sit — 11
4. Gedanken Spiegel — 15

Episode Two — 17

5. Who Knows What When? Information Processes in the EPR Experiment — 19
6. The Guy with the Red Flag Standing in the Middle of Nowhere: The Einstein-Podolsky-Rosen (EPR) Paradox — 21
7. EPR's Gedanken Experiment vs. Heisenberg's Uncertainty Principle — 23
8. Conspiracy vs. Knowledge — 27
9. Bell's Theorem: It's Easy — 29

10. Certainty vs. Uncertainty: Substratum of Chaotic
    Classical Forces vs. Quantum Mechanic's Probability   31
11. Randomness of What?   33
12. Separability and Inseparability: Out There vs.
    Nospacetime   37
13. The Tree of Life vs. the Quantum Leaps of Life   41
14. I Know What You Know: The Experimental
    EPR Outcome   45
15. The Clouded and the Unclouded Mind:
    A "Knows" about B   47
16. Local and Nonlocal Mindset   49
17. Creation of Knowledge Out of Nothing   51
18. The EPR Outcome in the Local Mindset   53
19. The EPR Outcome in the Nonlocal Mindset   55
20. Beyond Modern Science: There Are Two
    Perspectives—You and Not You   59
21. The Einstein-Podolsky-Rosen (EPR) Experiment:
    Turning Guns and Bullets into Roses   61
22. Einstein Is Right, Despite the Error in His Thinking   67
23. Space-Time Dependent Interpretation:
    The Local Mindset of Normal Science   69
24. Space-Time Independent Interpretation:
    The Nonlocal Mindset   73

Episode Three   77

25. Riding the Cannonball of Light: Formation
    of an Information Singularity at the Speed of Light   79
26. What Exists at the Speed of Light   83

Episode Four   87

27. Two Worlds That are One: The General Theory of Information   89
28. The General Theory of Information: Two Stories to Tell   91
29. Definition of the General Theory of Information: Don't Worry—You Can Use It!   93
30. Can We Understand the Phrase "Infinite Amount of Inseparable Information"?   97

Episode Five   99

31. What Now? Everything at Your Fingertips: Let's Work Together   101
32. Ex Nihilo Nihil Fit: Birth and the Mystery of Life   105
33. Ex Nihilo Nihil Fit—Nothing Comes From Nothing   107
34. Birth and the Mystery of Life   111
35. Unfolding Nonlocal Information Via the Human Genome: Complexity and Simplicity   113
36. Complexity and Simplicity   117

Episode Six   121

37. Who You Are—What You Have to Do   123
38. Waiting for Godot: Waiting for Somebody Else to Do Our Jobs   125
     Readings   127

THE GENERAL THEORY OF INFORMATION

Origin of Truth and Hope

John Archibald Wheeler summarized his life in physics as follows:

I think of my lifetime in physics as divided into three periods. In the first period ... I was in the grip of the idea that Everything is Particles ... I call my second period Everything is Fields ... Now I am in the grip of a new vision, that Everything is Information. [1]

The following book illustrates the profound significance of this vision.

....

I think it's much more interesting to live not knowing than to have answers which might be wrong.
—RICHARD P. FEYNMAN

....

It's 2012. It's time.

—BACH & BELARDO

....

Prologue

We are not small, insignificant pieces of a big machine. We must consider ourselves as having the ability and response-ability to build a world of justice, equality, and mutual prosperity.

This book is about who you are.

This book is about what you are.

This book is about two words: infinite and inseparable.

This book is about the unthinkable and unspeakable.

This book is about the unthinkable nature of who we are.

This book is about the unspeakable truth of us.

This book is about the hope that lies in our ability to eliminate hunger and poverty.

This book is about absolute knowledge and truth and about the unthinkable consequence of absolute responsibility.

This book is about the ONE consisting of the MANY.

This book is about the responsibility of the ONE.

This book is about the fact that inaction does not relieve one from responsibility.

The words do not make successful books, but rather the absolute meaning they convey between the lines.

The General Theory of Information is the certainty of hope. Hope is the driver of all action, entitling us to take matters into our own hands and build a just, stable, and prosperous world now. Only when our action is guided by the discovery of our true nature may it serve the common good. Our unique information-based perspective tackles the unspeakable and unthinkable in science by offering a step-by-step guide to understanding the truth about ourselves.

The General Theory of Information boldly describes two realities, two mindsets, two reference frames interwoven with one people, one responsibility, and one hope. It employs this "information perspective" as a common-sense approach to make the science of the unspeakable and unthinkable truth about ourselves accessible. The book is written in clear and understandable terms, offering the reader ready-to-know ideas that can be used for the highest good of all. Its purpose is to articulate the General Theory of Information and to define its two key concepts: "nonlocal information," meaning an "infinite amount of inseparable information," and "nonlocality," meaning "undividable inseparability."

The origin and scientific validity of the theory go back to the Einstein-Podolsky-Rosen (EPR)-experiment and the fact that time equals zero at the speed of light. The EPR experiment provides conclusive scientific proof of the fundamental inseparability of reality. An analysis of information processes in the EPR experiment reveals properties that serve as the basis of a novel nonlocal mindset.

A nonlocal paradigm allows thinking the unthinkable and speaking, or at least developing a language for, the unspeakable, beyond the paradigmatic scope of normal science, unimpeded by established concepts and unconstrained by local reality. Suspension of previous knowledge may be necessary to allow the growth of this new knowledge about nonlocal aspects of reality that lie outside of our experience of an observable local reality, and in a Kuhnian sense, outside of normal science.

It is 2012. It is time to do good, take responsibility, share love, and build paradise!

Preface

We want you to stop thinking that you are a small, insignificant piece of a big machine. Discover your own nature, take 100 percent responsibility, use your creativity, and find ways to change the world peacefully. Make peace with yourself and the world. Advances in science depend on new thinking and new mindsets, not so much on new experiments in old mindsets that are always producing the same paradoxes and violations of old paradigms.

—BACH & BELARDO

To raise new questions, new possibilities, to regard old problems from a new angle, requires creative imagination and marks real advance in science.

—ALBERT EINSTEIN

Hope Is the Driver of All Action

I learned very early the difference between knowing the name of something and knowing something.
— RICHARD P. FEYNMAN

The meaning of words might be as elusive as the wind. You cannot see wind; you can only observe a gentle, fleeting force weaving through a field and only for an infinitely small moment.

— BACH & BELARDO

Everybody else is supposed to come save us—except us.

— BACH & BELARDO

Hope is the driver of all action. Discovering the true nature of ourselves serves to guide our actions toward the good of self and other. The General Theory of Information is the certainty of hope, which entitles us to build a just, stable, and mutually prosperous world. This book, tackling the unspeakable and unthinkable in science, entitles you to stop thinking that you are a small, insignificant piece of a big machine. There is no machine, only you in the form of the unthinkable and unspeakable infinite many, not coming from nothing and going into nothing, but entitled, through your ability and responsibility, to build a world of justice and positive action, equality, and responsibility where everyone can prosper. It is this very idea that cannot be conceptualized nor thought or spoken of, and yet it exists.

Hope is the driving force behind all of our decisions, actions, and accomplishments that define the state of this world. It is the foundation for everyday decisions we must make: we must decide every day where we want to go, what the state of this world is. Hope means we take responsibility; we take matters into our own hands. In our time on planet Earth, we must envision a confluence of truth and hope into a virtual tsunami of humanity. Never before in history have we seen a noble world that serves the people. It is time to build and not destroy. It is time to do, and not merely demand.

The General Theory of Information furnishes proof that we are indeed ONE, with ONE responsibility and ONE hope, which implies we do not have to ask for outside consent, we simply consent to be doers ourselves.

How far can we get writing about the unspeakable and unthinkable? Probably not far enough to make the reading an easy walk. When we take this challenge to tackle the unspeakable and unthinkable, the goal is to arrive in a step-by-step fashion at clear and understandable terms and ready-to-know ideas that can be used for the highest good of all. The General Theory of Information describes two realities (local and nonlocal), two mindsets (local and nonlocal), and two reference frames (local and nonlocal). In this context, local means spacetime, and nonlocal means beyond spacetime (nospacetime) that is interwoven with one people, one responsibility, and one hope. It employs an "information perspective" as a common-sense approach to making the science of the unspeakable and unthinkable truth about ourselves accessible.

> The one word that is the catalytic agent in this context, and that will open our eyes and minds for seeing truth, is *inseparability*. Its origin lies in experimental science, its content is the unspeakable and unthinkable truth, and its meaning is hope.

In the first episode we give a brief account of the whole story: the "origin of inseparability" in the EPR experiment and the fact that at the speed of light, time and space equal zero, and mass, energy, and information become infinite. To avoid those unspeakable and unthinkable consequences of inseparability, the term "instant-action-at-a-distance" was introduced, which unfortunately prevents us from seeing the truth, and it induces a way of thinking that hinders progress in science. If we truly want to develop an understanding of the unspeakable and unthinkable, we have to know that a new mindset and new terms are needed. Because of the lack of language and words, we use the terms "nonlocal mindset" and "nospacetime."

It is important to understand that the new "nonlocal mindset" is complementary to the old "local mindset" and is designed to foster progress in science on both the local and nonlocal paths. Practically speaking, the nonlocal mindset does not exist, but it is necessary to make the unspeakable and unthinkable in science accessible. Thus, this book is an invitation to the reader to participate in the setting up of new nonlocal thinking that has the good of self and other as its focus.

The new nonlocal mindset, which needs to be strictly kept separate from any local mindset, takes into account the "inseparability of things," whereas any local mindset proposes "separability of things," that is to say, things, under certain conditions, exist independently of each other. Again, both paradigms exist and are complementary to each other and should not be used for arguing against the other. Any argument would eliminate constructive thinking, innovation, and progress in science and for humanity.

The term nospacetime—again a word used for lack of appropriate language—is a construct that is obvious when taking into account that at the speed of light time and space equal zero. This fact has been taken lightly by many scientists because of its undesired impact, according to Hugh Everett III, that it practically makes the mathematical tools of science impractical, and he goes on, "in no way is [it] evident to apply the conventional formalism of quantum mechanics to a system that is not subject to external observation. The whole interpretative scheme of that formalism rests upon the notion of external observation[2 p.455]" – a reality "out there". This applies to any mathematical formalism. So the problem with inseparability is that there is no external reality that can be observed by an external observer. While it is understandable that we would avoid dealing with the notion of an inseparable reality, we have to deal with the fact that inseparability is *the* truth, but at the same time separability is our truth, and both must be linked together without eliminating the other. So the point is not to strive to determine what is right or wrong, but to acknowledge that things are either right, or right. The point is not "the winner takes it all" but "we all are winners."

Thinking along this new path, the "nonlocal common sense" conclusion is that because light travels at the speed of light, it must exist in nospacetime and can still have an infinite mass. However, a popular argument for light traveling at the speed of light is that it has no mass. Thorough reflection on this matter reveals that it shuts down any constructive thinking. It stops people from thinking about what is really going on because they assume having no mass is the solution without further thought. The problem is that we don't know what mass is, and we stop thinking about mass when we are confronted with the fact that mass is infinite at the speed of light, and we unconsciously say,

"This cannot be true, though; light has no mass and that's it." Then we move on with our lives and careers, which is what Einstein did.

But, yes, light has infinite mass. The problem is not that light has infinite mass, but our thinking about it. Our way of thinking creates the problem, not the special theory of relativity that accurately states that the mass of any object, including light, is infinite at the speed of light. This book is ultimately about examining our unidirectional limited restrictive thinking, and about leaping over those restrictions easily with constructive reasoning.

In the exhaustive second episode, the new terms nonlocal mindset, nonlocal information, and nospacetime, among others, are further elaborated in the context where we describe the origin of our theory in the Einstein-Podolsky- Rosen-experiment,[3] which provides experimental evidence of nonlocality, not in the sense of instant-action-at-a-distance but simply as "true undividable inseparability" or "inseparability of reality." [4,5]

Our General Theory of Information adopts a fresh view for examining theories, phenomena, models, conventions, and experiments, which led us to the discovery and formulation of new information-related phenomena and processes. Nonlocality in a new nonlocal way can, therefore, be regarded as "instant information," "omnipresence of information," and "instant connectivity" in a local reality. Nonlocal information, on the other hand, exists at and beyond the speed of light where time and space do not exist. We describe the EPR experiment together with an explanation of the experimental origins of the two key concepts nonlocality and nonlocal information.

The third episode deals with the nature and behavior of time, space, and information at and beyond the speed of light. Our analysis incorporates the scientific convention that time equals zero at the speed of light[6]. In our information-oriented interpretation, this leads to the formulation of the two new concepts: "information singularity" and "nospacetime" or "spacetime singularity."

The underlying "inseparability of reality" is the key to understanding that when "local information from the past, present, and future collapses" into a "single point of nonlocal information" at the speed of light, the result is an "information singularity." Likewise, when time and space collapse at the speed of light, the result is the formation of nospacetime in a nonlocal sense—a "spacetime-singularity." Consequently, the information singularity needs to exist in nospacetime, which accounts for the inseparability of reality[7,8]. In this view, nonlocality requires nospacetime. The obvious nonlocal interpretation is that since nonlocal information is independent of spacetime and can exist beyond the speed of light, it forms an "inseparable information singularity at nospacetime." Furthermore, the finding that nonlocal information is an inseparable reality means that an observer would have to have an infinite information processing capability in order to observe reality as a whole. Consequently, any "limitation of the infinite amount of information" creates an "observer-dependent reality." Therefore, nonlocal reality remains invisible to observers whose capacity for information processing is limited.

In the fourth episode we provide a formulation of the General Theory of Information, accompanied by a description and definition of nonlocality and nonlocal information. Our theory defines reality as an inseparable nonlocal information system in which all information (past, present, and future) is instantaneously known everywhere. Information does not vanish at the speed of light, but instead it forms an information singularity in nospacetime. It follows from this that reality beyond the speed of light consists of an infinite amount of inseparable information, which cannot be grasped by any local spacetime dependent theory[9]. By limiting nonlocal infinite information that represents nonlocal reality through the act of our observation, we create any number of "local realities," which are composed of "local finite information."

The fifth and sixth episodes discuss fundamental practical implications of our theory on our individual daily thinking and acting as well as society at large.

EPISODE 1

When you know how others think, develop a new thinking.
—ALBERT EINSTEIN

Nonlocal thinking.
—BACH & BELARDO

The Whole Story

Anyone who has never made a mistake has never tried anything new.
— ALBERT EINSTEIN

The only thing we can agree on with certainty is that we cannot understand. This is the first step toward progress.

— BACH & BELARDO

The purpose of this book is to articulate a General Theory of Information and to define its two key concepts: nonlocal information, defined as "infinite amount of inseparable information," and nonlocality, defined as "truly undividable inseparability." The origin and scientific validity of the theory go back to the EPR experiment and the fact that time equals zero at the speed of light. This experiment provides conclusive scientific proof of the fundamental inseparability of reality. An analysis of information processes in the EPR experiment reveals properties that serve as the basis of a novel nonlocal mindset. A nonlocal paradigm allows for thinking the unthinkable and speaking, or developing a language for, the unspeakable, beyond the paradigmatic scope of normal science, unimpeded by established concepts and unconstrained by local reality. Suspension of previous knowledge may be necessary to allow the growth of new knowledge about nonlocal aspects of reality that lie outside of our experience of an observable local reality.

From a novel information-perspective, we may view experiments as information systems for the purpose of identifying and understanding phenomena and processes that have gone unrecognized because they lie outside the traditional scope of "normal science."[10] This book provides the foundation for a new science that can cope with an infinite number of paradigms so that we can conceptualize reality focused less on prediction and control and more on explanation and understanding. Paradigms are one way to cope with our severely limited intellectual capability, bringing the unspeakable and unthinkable to a level that is accessible for our limited capability to process information and be able to understand. We simply cannot rationally cope with infinity and inseparability. These two words alone shut down thinking and thus any hope for progress. Truth of the matter is that it takes an infinite number of conceptualizations to describe reality with the tools and formalisms we have developed.

A conceptualization is the question we ask, which is the experimental set-up and the experimental result we get. If we ask for the concept of position, we get a position, but no particle. If we ask for a "quantity of inertia" or "quantity of momentum," we produce only those "quantities of a force" without creating real physical objects per se.

We arrive at our definitions by interpreting the EPR experiment in a new light and by seeking to answer two questions, namely: is there experimental evidence that might provide a sense of what can be considered absolute scientific knowledge; and, does absolute knowledge let us understand the absolute truth about ourselves?

To reiterate, the EPR experiment does provide conclusive scientific proof of the fundamental inseparability of reality. Our new theory of information solves the paradox of nonlocality, which has been defined <u>erroneously</u> as instant-action-at-a-distance and requires information to travel faster than the speed of light. A reconsideration of the notion of instant-action-at-a-distance from a new information perspective leads us to conclude that the phenomenon of nonlocal information constitutes an inseparable nonlocal reality. Our General Theory of Information defines nonlocal (inseparable) reality as inseparable nonlocal information, which can be limited by finite information processes that generate local (separable) realities. Nonlocal information is defined as an "infinite amount of inseparable information." Nonlocality is thus defined as inseparability, or more precisely, the inseparability of reality. Our theory is intended to complement rather than replace existing scientific theories and frameworks. We call attention to the limitations imposed by language and preconceived ideas regarding reality, absolute knowledge, and truth.

Bell's Speakable and Unspeakable

(Joan, 1941): She wrote me a letter asking 'How can I read it? It's so hard.' I told her to start at the beginning and read as far as you can get until you're lost. Then start again at the beginning and keep working through until you can understand the whole book. And that's what she did.
—RICHARD P. FEYNMAN

Think before, during, and after reading. But separate new nonlocal thinking from all you have learned so far. With logic one can create some knowing, but only wisdom and intuition lead to understanding and knowledge. There are limitless ways of knowing and knowledge, and one truth can be divided into an infinite number of truths.
—BACH & BELARDO

Everybody keeps complaining that they cannot understand anything. As a start you might commence with thinking by yourself and not waiting for somebody to explain something that

most likely is leading away from truth. Not fooling yourself and seeing through lies and misconceptions is the real challenge.
—BACH & BELARDO

"Sleepwalkers" is the word J.S. Bell[11 p. 169] uses in his book *Speakable and Unspeakable in Quantum Mechanics* to describe our attempts at making sense of our experimental outcomes and what they mean. We are more sleepwalkers than people of wisdom.

Unspeakable is that light has infinite mass at the speed of light. Speakable is that light has no mass, but this does not explain anything, nor does it provide a solution. It just shuts down thinking and reflection, because everybody assumes the case is solved. Nothing has been solved. We must think the unthinkable—that light indeed has infinite mass at the speed of light. Now what is the thinkable solution? You might need to sleepwalk to establish an infinite number of solutions. Rational thinking won't produce even one genuinely new idea.

Our line of reasoning is based on the observation that the EPR experiment seems to allow instant-action-at-a-distance, and, thus, that *information travels faster than the speed of light*. Since existing theories cannot fully explain this phenomenon, it is necessary to look at instant-action-at-a-distance (events affecting each other instantly) and related notions from a different, information-based perspective and reevaluate existing conventions and experiments[11,12]. The following inquiry focuses on information phenomena and processes that are evident in modern scientific experiments but lie outside the traditional scope of normal scientific inquiry. We employ a radically new perspective from which we may reevaluate experiments, observations, and the world in general. At first glance, our methods may appear surprisingly simple or even trivial. However, our theory has far-reaching implications, because it provides a coherent explanation of the nature and origin of reality and develops new and unique concepts.

We use the example of light to illustrate properties that exist beyond space and time. When traveling at the speed of light, time and space do not exist. Imagine "distance" as a straight line between the sun and the earth and a light-photon traveling along that line at the speed of light. Now envision the light-photon as a cannonball that you are riding. Your perspective would differ from that of an observer on earth. The observer on earth sees the cannonball crossing the distance along a straight line, which requires time. From your cannonball perspective, however, things look quite different. The cannonball on which you are sitting does not have the properties of *speed* as quantity of motion arising from the velocity and quantity of matter conjointly (momentum) and *location*. It does not seem to be traveling at all, because the distance between the sun and the earth is created by the observer's measurement of movement relative to the position of observation on earth.

Unfortunately, this thought experiment is subject to the severe limitations of language. The notions that "because time is zero, it takes zero time to get somewhere" and "because space is zero, there is no distance" are descriptions of local events (local, meaning at a point in spacetime vs. nonlocal, meaning no-spacetime). Here, they are used to describe nonlocal processes in a nonlocal reality where spacetime does not exist. The words somewhere, time, space, and distance are meaningless, because in a nonlocal reality, everything is in one place. Similarly, place has no meaning in a nonlocal reality, and there are no words to express nonlocal nature [13].

A Singularity without a Place to Sit

The first principle is that you must not fool yourself, and you are the easiest person to fool.
—RICHARD P. FEYNMAN

Genuine new ideas might knock on your door seeking entrance. Let them in and progress might show itself. Don't fool yourself with old thinking; be open to new ways of thinking.
—BACH & BELARDO

With the Big Bang all matter was collapsed into one single point before space and time existed. Yet how can we define this "one single point" or know where it took place? Does the word "singularity" help clarify the nonlocal nature of reality? We may not know what it *is*, but we do know what it *is not:* a singularity is not a point and has no location; it is everlasting and infinite, with no beginning or end. A "singularity" simply *"is"* and contains "everything."

Ultimately, however, a singularity is precisely what *has* been observed in the EPR experiment: "There is no distance;" thus,

"There is no instant-action-at-a-distance." Light itself proves that there is a reality beyond space and time, which may be termed nonlocal reality. Local and nonlocal reality are complementary to each other, and light exists in both. Light can be observed by an observer on earth in a local reality in spacetime, but at the same time exists as a nonlocal entity in a nonlocal reality beyond spacetime. This "dualism of properties" of light cannot be fully explained by conventional theories, and the explanation of these unobservable phenomena lies outside the traditional scope of physics[10]. Information, similarly, is unobservable, but it is created by a mind and has meaning only at the particular moment the meaning has been created; a moment later, the meaning will have changed. For example, every time we use the word car, even a second later, we envision a car slightly different, and we don't even remember our exact thoughts the last time we used the word car—what color, what make, etc. There is not one word that can be defined uniformly that retains the exact meaning for everyone.

In physics, the concepts of observable information and information processes are limited to signals and do not refer to "meaningful" information per se; they apply to transmission of meaningless data pieces as defined by Shannon[14], who developed mathematical rules for communication but did not address the nature of information itself. Meaning, per se, is created by the sender and/or recipient, but it is not an intrinsic value of a signal. If there is no recipient, any meaning intended by the sender is meaningless. Although the EPR experiment corroborates the idea that information travels faster than the speed of light and thus establishes the notion of nonlocality, the explanation and understanding of nonlocality in traditional physics is limited to interpretations that remain in the realm of local theories and assign properties to nonlocality that can be observed and described within the confines of the spacetime continuum.

In contrast, our novel information-based perspective defines nonlocality ontologically as having the properties of a nonlocal reality independent of spacetime. Because the new theory arrives at an explanation and understanding of nonlocality that is different from other disciplines, it is *sui generis*. The theory presented here serves as the scientific and theoretical foundation for a novel science that recognizes nonlocal reality on an ontological level. It may also be used to facilitate interdisciplinary collaboration and cross-fertilization. The goal is to establish a common ground, mediate differences, and provide an intellectual platform for the integration of the unique perspectives of various academic disciplines.

GEDANKEN SPIEGEL

If you can't explain it simply, you don't understand it well enough.

—ALBERT EINSTEIN

Believing everything that is believable is not a panacea for new thinking. Free your mind and open yourself up for the unbelievable and the unthinkable.

—BACH & BELARDO

Because of its focused scope, this book serves as an introductory *Gedanken Spiegel* or *Gedanken* experiment; the objective is to mirror existing thoughts in a new, more coherent and complete framework. The scope of our investigation is limited to one experiment and one theory of modern science. The conclusions are not necessarily the only possible ones and are intended to show new ways of investigating and reasoning that can lead to a creative dialogue between scholars from various disciplines.

Our unique information-based reasoning produces a General Theory of Information where reality is defined as a spacetime-independent information system in which finite information processes generate local, observable phenomena. This new understanding allows us to identify and explain ill-defined phenomena that do not fall within the purview of existing disciplines.

Episode 2

One thus sees that a new kind of theory is needed that drops these basic commitments and at most recovers some essential features of the older theories as abstract forms derived from a deeper reality in which what prevails in unbroken wholeness.
—DAVID BOHM

Old theories are a fertile ground for new discoveries if you start with no commitment and develop many ways of new thinking.
—BACH & BELARDO

WHO KNOWS WHAT WHEN? INFORMATION PROCESSES IN THE EPR EXPERIMENT

... in the experiments about atomic events, we have to do with things and facts, with phenomena that are just as real as any phenomena in daily life. But the atoms or the elementary particles are not as real; they form a world of potentialities or possibilities rather than one of things or facts.
—WERNER HEISENBERG

Truth might have unintended consequences. For instance, it might not be what we expected.
—BACH & BELARDO

The aim of this chapter is to describe the EPR experiment, to discuss the implications of the definition of nonlocality and instant-action-at-a-distance, and to examine the phenomenon of nonlocal information. Nonlocality is an aspect of reality that is counterintuitive to our understanding of the world. Its converse, locality, is the principle that says an event at one place cannot affect an event somewhere else simultaneously. For example, if a supernova explodes today, the principle of locality stipulates that an observer on earth could not possibly know about or be affected by this event until, for example, a photon traveled to earth.

The Guy with the Red Flag Standing in the Middle of Nowhere: The Einstein-Podolsky-Rosen (EPR) Paradox

How wonderful that we have met with a paradox. Now we have some hope of making progress.

—NIELS BOHR

A paradox is when you know that you don't even know that you don't know.

—BACH & BELARDO

When you stand in the middle of nowhere, you are everywhere—simultaneously. This is what you can know but not understand. In a 1935 article[15], Einstein, Podolsky, and Rosen (EPR) published a "Gedanken experiment" showing that, under certain circumstances, quantum mechanics predicts a breakdown of locality and that measurements of or at systems separated in space can influence one another instantaneously, with no lapse of time. The assumption that locality is an underlying principle of reality not only seems intuitively correct but is also supported by theoretical—but not practical—implications of the special theory of relativity, which predicts that nothing can reach the speed of light, because, according to $E=mc^2$, mass would be infinite at the speed of light, and no signal, therefore, can be transmitted faster than the speed of light.

EPR's Gedanken Experiment vs. Heisenberg's Uncertainty Principle

No, no, you're not thinking; you're just being logical.
—NIELS BOHR

Logic starts with a set of easy rules and builds a web of reasoning from which there is no escape and only limited progress.
—BACH & BELARDO

Some say it is a waste reading about truth, because there is none, meaning: don't think—thinking is a waste. Be a follower.
—BACH & BELARDO

Theorists assumed the locality argument to be irrefutable and concluded that quantum mechanics was incomplete and/or flawed. The basic principle of EPR's Gedanken experiment contradicts Heisenberg's uncertainty principle (you cannot know two properties of something at the same time: e.g., if you see that a car is blue, you cannot know its horsepower or other properties), since it seems to prove that both position and momentum (speed) of a particle *can* be known simultaneously (the solution is that there is no car or object, only the perception of blue by a limited observer who cannot see ergo process information that is more than blue).

Einstein, Podolsky, and Rosen conceived an experiment in which an entangled pair of particles (A and B) with equal mass and opposite momentum is split up and set flying in opposite directions. Because one particle has a positive or up momentum while the other particle has a negative or down momentum, Einstein, Podolsky, and Rosen constructed a wave function that allowed them to deduce both momentum and position of particle A by measuring particle B and thereby *gaining knowledge* about particle A without disturbing or observing it. On this basis they argued that since it is possible to predict with certainty the value of a physical quantity (momentum and position) without disturbing the system, there is a property of physical reality corresponding to the value of the physical quantity. In other words, Heisenberg's uncertainty principle is violated because the momentum and position of both particles can be known at the same time.

The experimenters' argument is that if A and B have flown a substantial distance away from each other, it is implausible that a measurement (observation) carried out on particle B at position B can affect particle A at position A. Particle A cannot be affected instantaneously, because that would be incompatible with the special theory of relativity's prediction that no signal of any kind can travel faster than the speed of light. Accordingly, it would be impossible for somebody at position A to know instantaneously (without any time lapse) that a measurement has been performed at position B. In our local reality, if one wants to use the remote control to switch on the TV, the signal from the remote to the TV always needs time, and one could know that the remote must have been used by observing that the "TV is on," without actually observing that the ON button on the remote was pressed. Therefore the property "pressing the ON button" can be known and would thus violate Heisenberg's uncertainty principle, which assumes that both "TV on" and "pressing the button" cannot be known simultaneously.

Conspiracy vs. Knowledge

We are all agreed that your theory is crazy. The question that divides us is whether it is crazy enough to have a chance of being correct. My own feeling is that it is not crazy enough.

—NIELS BOHR

There is a local and a nonlocal reality or truth or theory. Beyond right or wrong—or crazy.

—BACH & BELARDO

Bohr rejected Einstein's reasoning and argued that although "no actual physical or mechanical force is transmitted" between A and B, the two particles seem to cooperate in a "sort of conspiracy" by *knowing* of each other's properties "at a given time." Bohr argued that it is meaningless to ascribe a "complete set of attributes" to any quantum object prior to measurement. Clearly, a practical experimental test was needed to distinguish between Bohr's and Einstein's views and test the locality argument of reality. It was not until nearly half a century later that such a test was developed.

Einstein considered absurd the notion that two widely separated particles could communicate so as to yield coordinated results of apparently independent measurements (momentum and position). He wanted his objective reality to be localized at each particle, and it was this locality argument that came into conflict with quantum mechanics[16].

Bell's Theorem: It's Easy

There are trivial truths and the great truths. The opposite of a trivial truth is plainly false. The opposite of a great truth is also true.
—NIELS BOHR

No way is evident to apply the conventional formulation of quantum mechanics to a system that is not subject to external observation. The whole interpretive scheme of that formalism rests upon the notion of external observation.
—HUGH EVERETT III

Two opposite but complementary mindsets. Two realities that are One. One truth that is many.
—BACH & BELARDO

In 1964, Bell studied the problem of two-particle quantum systems and formulated a mathematical theorem that turned out to be of crucial importance in setting up a practical EPR experiment[17]. He investigated the possible correlations between measurements conducted simultaneously on two entangled then separated particles and established the theoretical limits of the extent to which the results of such measurements can be correlated. This theorem provided the conceptual framework within which expected values can be calculated for the spin components of a pair of spin-half particles—one particle has a spin of plus half (+1/2) and the other a spin of minus half (-1/2)—then showing that, regardless of the choices made for certain unspecified functions occurring in the framework, the expectation values obey a certain inequality. This is known as Bell's Inequality, which is essentially independent of the nature of both particles or the details of the forces that act on them, focusing instead on the rule of logic that governs all measurements. For example, a population census of a multicultural society cannot possibly find that the number of Caucasians exceeds the number of white males, plus the number of females of all races[16].

Certainty vs. Uncertainty: Substratum of Chaotic Classical Forces vs. Quantum Mechanic's Probability

When it comes to atoms, language can be used only as in poetry. The poet, too, is not nearly so concerned with describing facts as with creating images.

—NIELS BOHR

Quantum behaviour is a product of a substratum of chaotic classical forces.

—ALBERT EINSTEIN

All observable facts are images. There is only one unobservable fact: inseparable infinity. The substratum of classical forces consists of an infinite number of classical forces: observable and unobservable.

—BACH & BELARDO

Let us assume that Einstein was correct in arguing that quantum behavior is the product of a "substratum of chaotic classical forces"[16 p. 16] that quantum mechanics is a theoretical framework and not an experimental one and that the practical implication of quantum mechanics is the prediction of measurement at a certain location at a specific time. Quantum mechanics does not explain what is measured or what the nature of the measurement is, nor does it have an explanatory framework to explain what the "object" or something like a particle is on which the measurement apparently has been performed. Prediction by itself does not necessarily imply the ability to explain the nature of what has been predicted. Relativity today is an experimental fact that predicts the limits of space and time (both equal zero at the speed of light) and what happens to classical forces relative to an observer. For example, the faster a visible macroscopic object moves, the more its mass increases until it reaches infinite mass at the speed of light, which again is a prediction and not an explanation. This raises the question of what Einstein's classical forces or phenomena are, what happens to them at or beyond the speed of light, and what their relevance is, assuming that classical forces are "elements of the physical reality" to which a "value of a physical quantity"[15 p. 777] can be ascribed.

If we were to apply "constructive reasoning" in which we do not question whether a thought of reasoning is right or wrong, we have at least two possible interpretations: In the first, we ignore what happens to classical attributes and carry on as usual. In the second, we face the fact that classical attributes become either infinite or zero at the speed of light: mass, energy, momentum = infinite; and time, space, position, spin = zero, which now leaves us struggling with the unthinkable and unspeakable meaning and truth of it all and acknowledging that we don't have the intellectual capacity to understand it, nor the language to express it.

The answer to these questions lies outside the theoretical frameworks of both quantum mechanics and the special theory of relativity.

Randomness of What?

God doesn't play dice!
—ALBERT EINSTEIN

Einstein, stop telling God what to do!
—NIELS BOHR

Believing or knowing? Either you know it, or you don't know it. If there are no real atoms, waves, strings, and objects, what should God play dice with?
—BACH & BELARDO

In Einstein's philosophy, the universe is a collection of assumed independently existing "objects" on which classical forces act that together make up the totality of a changing universe. Unfortunately, Einstein did not explain the significance of the various forms or varieties of objects, e.g., particle, wave, string, tachyon, or meson. Einstein's philosophy, moreover, requires that these objects exist in space and time "at all times," despite the facts that reaching the speed of light is prohibited for any of them and no signal of any kind can travel at or faster than the speed of light.

Appropriately formulated, Einstein's first assumption is that the "real nature of objects" is to consist of a complete and finite set of "measurable properties of classical forces" that are of such kind as mass, momentum, spin, and place, all of them possessing exact and immutable values. For example, spin possesses either the absolute value of plus or minus, and momentum has an absolute value of velocity at a certain position in space, even though we cannot measure position and momentum simultaneously (uncertainty principle).

There is no indication that we can know how many properties of classical forces do constitute a complete and finite set so as to constitute physical reality, and challenges remain to account for phenomena like these: for mass to possess an immutable value, time and space must be absolute so that the mass does not change relative to speed and energy. Most importantly, however, is the notion that "there can only be" one predefined observable reality "out there," and that there is no choice of other parallel forms of reality.

In summary, Einstein's prime assumption is that objects really do possess quantitative immutable properties in a well-defined absolute sense at all times and at each point on a trajectory from A to B.

We must not, however, ignore the fact that for objects to be "real at all times in space" entails "the presence of a nonvanishing spatiotemporal interval,"[4 p.226] meaning that there can be no condition in which time and space do not exist (equal zero), which would mean that time and space at one point remain absolute (nonvanishing). Because time goes slower and space shrinks when increasing the velocity of an object, this would mean that nothing can reach the speed of light, which can be falsified with our ordinary observation simply by observing "something" traveling at the speed of light.

Einstein's second assumption is termed "locality" or "separability," because it forbids objects from instantaneously exerting physical influences on each other when they are separated by space and time. They simply cannot occupy the same location at the same time and so need to occupy different "stretches of space." In this context, a "local" position has the attributes of being "separated" and "independent" (not connected), thus allowing us to assume the existence of a "local reality." On the basis of this assumption, Bell was able to establish a strict limit on the possible level of correlation for the simultaneous measurement of two particles. Bell's Theorem highlighted the conflict between Einstein's notion of a local world composed of independent objects and Bohr's ghostly world of "subatomic conspiracies."[16]

Einstein's argument about locality would be correct if the experimental results did not violate Bell's inequalities. Any violation would prove the converse, namely, nonlocality, which means the inseparability of reality. Picturing the world as full of independent objects is compelling to those of us conditioned by Western philosophy and conforms to our common sense understanding of what nature is supposed to be, however vague that may be. Einstein called it "objective reality," because objects have independent properties and exist in a universe that does not depend on an individual's observation or an unconscious measuring apparatus (a click on the counter). It was precisely this common-sense view of reality that Bohr challenged, arguing that things do not exist before they have been observed or measured.

It still is unimaginable that things do not exist even after they have been observed, or that things have a nature that is unthinkable. So, does an object exist after measurement? No, only the measure exists: of mass, momentum, position, or any other observable classical force, but there is no object that could possess or exert an additional classical force simultaneously with the measurement. For example, if we measure the momentum, we cannot measure the position, "because there is no object." By the same token, if we measure the position, we cannot measure the momentum, because there is no object.

Again, we don't have the intellectual capability to understand this. What we can know is that on a subatomic level there is a "nonlocal reality" with no space and no time, thus no objects with classical forces of any kind, but there are still properties and forces that exist in the nonlocal reality that we can see and measure in our local reality. For example, we can see light even though it exists in nospacetime, and there are some unknown forces that do not exist in space and time but that are responsible for accelerating the expansion of the universe. This leaves us with the challenge to develop some sort of understanding of what such "nonlocalizable properties and forces" are. With quantum mechanics we have the formalism with which we can predict the emergence of "nonlocal properties and forces" in our local reality, even though we have no understanding of them.

In contrast with the nonlocal world, we have our local world in which we can measure certain classical forces of planets, bullets, and water waves and can give them properties related to space and time that do exist with certain positions and momentums at each point on a trajectory between two locations.

Separability and Inseparability: Out There vs. Nospacetime

If anybody says he can think about quantum physics without getting giddy, that only shows he has not understood the first thing about it.
—NIELS BOHR

At the speed of light, time and space cease to exist and with it the notion of something being "out there."
—BACH & BELARDO

We can know that there is something unthinkable, like nospacetime (at the speed of light), but we cannot understand what nospacetime is. That's how far we can get in expressing the unspeakable and unthinkable in this book. It now puts us before the task to develop an understanding of the unconceptualizable concept of nospacetime. We cannot just say that there is no time and no space, because then there is no progress in thinking. We have to admit that there is something at the speed of light, even if we cannot express and give form to this still unthinkable reality.

To speak the unspeakable and to think the unthinkable might require taking a detour in thinking. Thinking about the prerequisites for something being out there might spark new ideas. For one, in order for something to be out there, time and space must be absolute; otherwise, there exists a condition where time and space cease to exist. The fact that at the speed of light space and time cease to exist results in the unthinkable circumstance that, simply speaking, literally nothing can be "out there" at the speed of light. Secondly, using another perspective, in order for something to be out there, it must be separate from an observer. In stark contrast to such local arguments, when you are nonlocal, you are here and there and everywhere simultaneously, but there is no location—just nospacetime, whatever it is. You cannot observe something in the out there. Again, you can know it but not understand it.

Before the emergence of quantum mechanics, Western philosophers and scientists usually assumed that the world around us had an independent existence, i.e., that it consists of enduring physical objects that are simply out there, whether we observe them or not. Likewise, any physical process described by a local spacetime-dependent theory will conform to the principle of spatiotemporal separability, which means that objects exist independently of each other and remain unaffected by sufficiently distant events. Such local theories assign properties to particles at each point along their trajectories through space and time[9]. Quantum mechanics, on the other hand, posits that objects *do not* move continuously through spacetime and *do not* exist at each point of their trajectory through space.

For example, electrons are thought to orbit the atomic nucleus (protons and neutrons) in distinct concrete orbits from which they cannot escape. If an electron moves from one orbit to another, they do a so-called quantum leap. Quantum leap indicates that the electron appears at Orbit B without having travelled along a trajectory and distance between Orbit A and Orbit B (arrow in Picture 1).

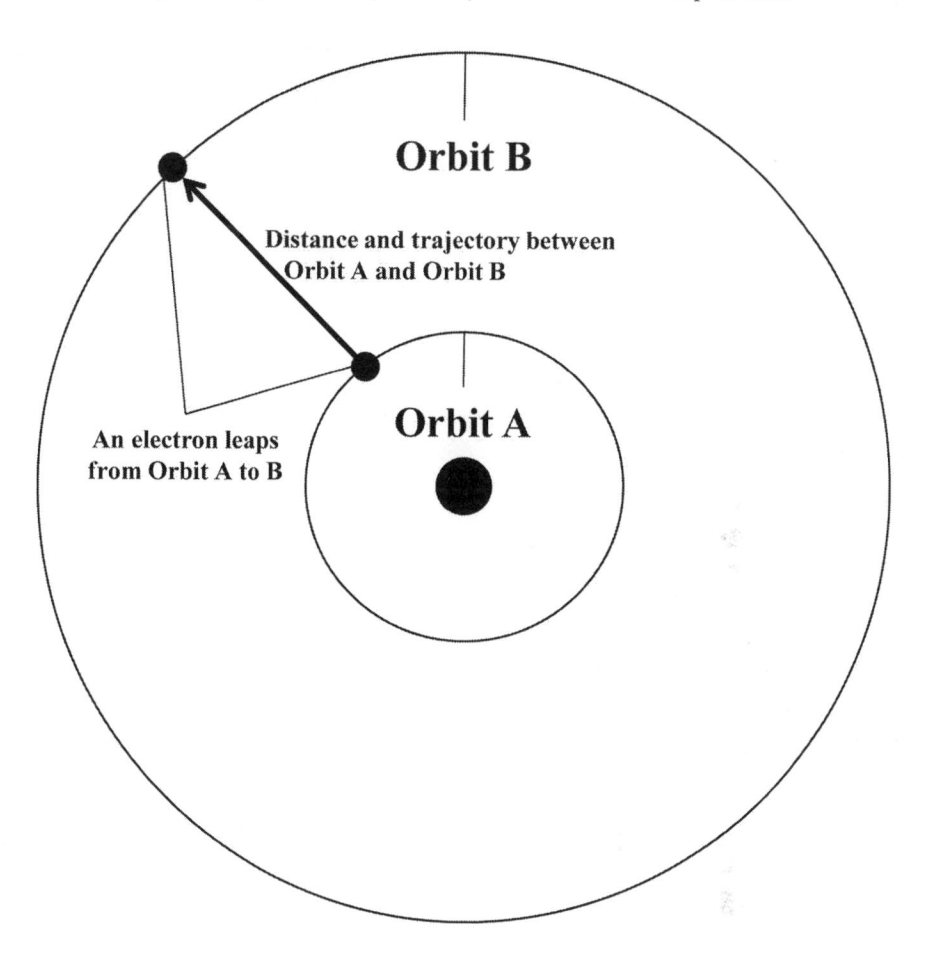

Picture 1: Quantum Leap of an Electron from Orbit A to Orbit B.

The electron vanishes from Orbit A and appears at Orbit B, seemingly leaping over the distance. This indicates that there is no space and no time and no gravity on the subatomic level. The electron would in a sense—assuming it exists as local entity at a location on Orbit A—disappear by switching from a local into a "nonlocal form of existence" and reappear as a local entity at a particular location on Orbit B without the requirement of crossing a spacetime interval.

Quantum mechanics does not describe reality, because there is no reality out there, and reality cannot be described or observed at each infinitely small point on a trajectory. Thus, the Bohr-Einstein debate about whether quantum mechanics describes reality completely or incompletely was beside the point, because quantum mechanics has a theoretical framework that predicts events and is not a descriptive framework that describes reality or objects therein. Quantum mechanics is a set of probabilistic rules and does not provide insight into any fundamental underlying truths. Because there is no reality "out there," there is no probability that can be ascribed to it. Life and consciousness do not depend on matter and are not created by matter since matter does not actually exist. Instead, matter is created by the limited information processing capacity of life, whatever life *is,* or *means*; thus, it is not matter that creates life but life that creates matter.

To resolve this debate, it was deemed necessary to test the fundamental assumption of separability, i.e., the notion that there is a real physical distance between things and that this distance is separated by a time lapse. The fact that it takes time to cross a distance is the fundamental essence of Einstein's locality argument. Bell's Theorem was designed to test this argument and dramatically changed scientific perceptions of reality. With the advent of quantum mechanics and Bell's Theorem, it became possible to discern the truth about reality without actually observing it. This allowed scientists to move beyond mere prediction and control to understanding and explanation.[18] In the following chapter, we discuss an idealized version of the EPR experiment to explain the central concepts of our theory.

THE TREE OF LIFE VS. THE QUANTUM LEAPS OF LIFE

Despite the fact that truth might be considered absurd in certain ways of thinking, truth still remains the truth.
—BACH & BELARDO

If, at first, the idea is not absurd, then there is no hope for it.
—ALBERT EINSTEIN

Just because something is unthinkable does not mean that it is wrong.
—BACH & BELARDO

The emergence of life is somehow linked to the idea of life forming out of disorganized matter that lumps together and ultimately begets life. Wow! Let's take this back one step and start thinking from the absolute beginning. The probability that life emerges from matter is quite challenging—beyond any probable causal cause for the existence of that matter. If we were to look at this from a nonlocal information point of view, however, things can be construed differently. Let's take the quantum leap way: The important point is that the emergence of the information for Species B (see Picture 2) is not produced gradually over time, but, in an imperfect sense, leaps, like an electron from Orbit A to Orbit B, in which the blueprint of Species B appears at evolutionary time B "abruptly," without traveling on a timeline from Species A. Some or most (> 99 percent) of the information of Species A might be contained in Species B, but this has no causal meaning—species B does not emerge because Species A was there. It can emerge anyway, with or without Species A. Meaning the unthinkable: there is no need for a local cause!

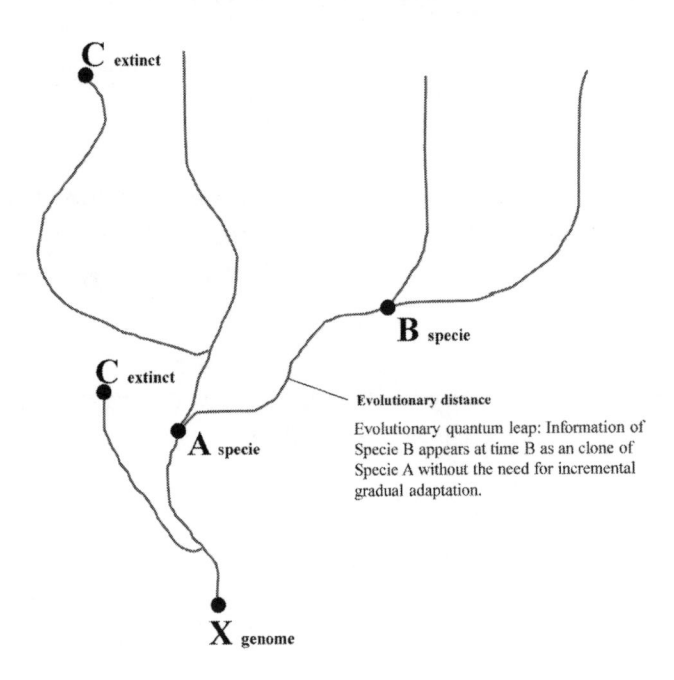

C extinct

C extinct

B specie

A specie

Evolutionary distance

Evolutionary quantum leap: Information of Specie B appears at time B as an clone of Specie A without the need for incremental gradual adaptation.

X genome

Picture 2: Evolutionary Quantum Leap. The information to create Species B at time B does not incrementally emerge from the information pool of Species A at time A but emerges at time B instantly, as proposed by the EPR experiment. Similarly, the information of the genome at X leaps into a local reality, and the information of extinct species at C in an unthinkable sense leaps into the nonlocal reality. The evolutionary distance is not a timeline with distinct points where at each of them slight changes in the genome occur.

The number of genetic changes needed in the blueprint of Species A to produce Species B do not need to emerge or be developed gradually in the spacetime continuum but can appear suddenly at time B without crossing the evolutionary distance, which would require making tiny genetic alterations at each point on a timeline. Small adaptations, for example the change of a mouse's fur color from light to dark, might follow time-related mechanisms that only require activating or deactivating a switch, but that do not require developing or randomly altering an entire gene. Again, existing theories should not be replaced but rather supplemented with nonlocal concepts that might give us a better understanding.

Similarly, the information of the X-genome does not need to be created out of nothing by random aggregation of matter but leaps into a local reality. We are not able to fully understand the how and why of this unthinkable phenomenon, but the mathematical improbability that billions of base pairs are randomly or semi-randomly aggregated has to be taken into account. In a nonlocal mindset we can view the information of the X-genome in the same way as we could imagine light as existing in a spacetime independent reality at the speed of light. Even though we can see light, it remains in the nonlocal reality of nospacetime where it retains unthinkable nonlocal properties such as infinite mass and traveling at the speed of light.

Now comes the unthinkable and unspeakable part where light leaps into our local reality and becomes able to perform an action, e.g. collide with an electron (photoelectric effect). Equally, the nonlocal information of the X-genome leaps into our local reality. However, the X-genome is not a local entity with a finite set of local information—genes and proteins—but represents an *apparatus mechanicus* between the local and nonlocal reality in which the genome translates nonlocal information into local action. The timing of local actions remains a nonlocal property. Likewise, the information of the C-genomes in an unthinkable sense leap into the nonlocal reality and is, in a more speakable sense, getting archived in an infinite library. The information does not vanish but leaps beyond our local reference frame.

I Know What You Know:
The Experimental EPR Outcome

The ability to perceive or think differently is more important than the knowledge gained.
—DAVID BOHM

Knowledge is created by a particular way of thinking. New ways of thinking create new knowledge.
—BACH & BELARDO

In 1982, Alain Aspect and colleagues challenged Bell's inequalities, rejected locality (but not Einstein's locality argument), and posited its converse, nonlocality, as the fundamental principle governing reality. However, the problem associated with defining nonlocality as allowing instant-action-at-a-distance is that the assumption of two independent systems, A and B, with some "distance" between them is an inherently local interpretation. Since the language of science has generally been based on the concept of locality, it is difficult to explain the EPR result. It is one thing to say that events affect each other instantly, but how an event "here" can instantly "know" what happens millions of light-years away, is not easy to fathom. For example, does a dog know instantly when its owner decides to come home, or was some prior information sent to the dog that requires time? The EPR experiment says, no, the dog did not receive any information it did not already have. The dog just knows instantly when its owner decides to go home. No information was sent, but the action or intention of the owner instantly impacts the entire universe, or in a nonlocal sense, picks the one local universe from an infinite number of universes.

For over half a century, physicists have been trying to understand this phenomenon and still have not achieved consensus. However, scientists generally agree that the results are valid and that something expressed as instant-action-at-a-distance is a fact, despite the challenge it poses to Einstein's argument that the speed of light is the speed limit of nature and that there must be a time-lapse to overcome whatever small distance. In the end, however, Einstein is proven right.

The Clouded and the Unclouded Mind: A "Knows" about B

Every sentence I utter must be understood not as an affirmation but as a question.

—NIELS BOHR

This book offers no ultimate understanding; it just invites you, the reader, to understand the challenge that there are an infinite number of directions to go.

—BACH & BELARDO

The words "knowledge" and "know" are key in how the EPR outcome can be looked at in different ways.

Local and Nonlocal Mindset

Even when you don't understand, you make paradise happen.
— BACH & BELARDO

A (the dog) "knows" about B (its owner) without receiving any information. A just knows; the information already existed at A. In the absence of appropriate language, A knows what happens at B before B knows what will happen to itself. This paradox arises because A and B are in different reference frames and conditions. In an unobserved (nonlocal) state, A remains in a nonlocal reference frame and has all information. When observed, B has a local reference frame, but only for the one observer—*not in the nonlocal world*. In reality, B is local and nonlocal at the same time. Only for the observer does B become local, finite, and distinct.

Our clouded mind, which is incapable of processing reality in its entirety, must think constructively in these two opposing local and nonlocal mindsets. It must recognize both their outcome as appropriate beyond the notion of it being right or wrong. This is beyond the grasp of our reasoning and understanding. An unclouded mind, in turn, can think in both mindsets by keeping them strictly apart.

CREATION OF KNOWLEDGE OUT OF NOTHING

Discussing truth is just shuffling words. It is like philosophy—just talking and no doing. Do Good! Share Love! Take Responsibility! Build Paradise!

—BACH & BELARDO

A's knowledge about B is like someone having a genuine ad hoc idea, whereby conceiving the idea would not require the notion of information being sent to or received by a mind or brain and without the need of having an indication as to where the idea might have come from.

We somehow just accept that an idea is created out of nothing. In the EPR case, A's knowledge cannot be created by associating previous information and combining it in a new way to create new knowledge. It is a genuine new idea.

The following are thinkable forms of how knowledge can be created out of nothing:

I. There is nothing before nothing. Creation occurs out of absolute nothingness. The first knowledge would be a word out of nothingness.

II. Creation by association. A new idea is created by combining known elements in a new way. The new idea or its information did not exist in this form before; there was nothing before the idea came into being.
III. Ad hoc creation. An idea just occurs. No relation to existing information or knowledge.

THE EPR OUTCOME IN THE LOCAL MINDSET

We don't have the language, we don't have the words, and we don't have the mindset.

—BACH & BELARDO

Science invented the external observer and never corrected that error. Similarly, science invented the concept of "right or wrong" and never corrected that error either.

—BACH & BELARDO

In the local mindset, A and B are separated. Einstein rightly argues that it takes time, under any circumstances, to send information from B to A. The problem in the local mindset is that A "knows" instantly, at the exact moment, when something happens at B, but A cannot know "before," only "simultaneously." The local mindset requires that a signal or a piece of information must be physically transmitted from B and has to arrive physically at A.

For example, B (Bob) is sending a birthday card to A (Alice). The birthday card and the wish are the physical information that travels from B to A. Even though Alice expects a birthday card from Bob, the card is sent at a specific time and needs to travel to Alice. This notion is based on two fundamental assumptions: (1) external observation: Alice and Bob are separate and completely independent entities, if there is a significant distance between them, and (2) Alice and Bob are finite entities containing different, limited, and separated sets of information.

In a nonlocal mindset, there is no distance. Bob is Alice (unthinkable and unspeakable). In a cautiously local sense, both consist of the same infinite amount of inseparable information, even though we perceive them as two individual entities, which isn't wrong, but you have to take into account that the nonlocal, inseparable bond, the "underlying inseparable nature," of them cannot be broken.

Local knowledge depends on some entity that knows.

In the local mindset, the local entity (A) has to know that something has arrived and then has to act on it. It cannot act or know before a piece of information arrives. It is the concept of a limited local mind.

Local thinking requires that a signal or information is something physical (birthday card) that travels from B to A. Local information has a physical, finite, and limited existence and is distinct, independent, and completely separated. When arriving at A, physical information requires a physical response to initiate an action.

When local knowledge is received at A, it induces action or reaction to something that is, from A's local perspective, created "out of nothing," because this particular knowledge did not exist before a signal or information has arrived.

In a local mindset, therefore, we must conclude that any transmission of information over a distance requires time. This has not been confirmed by the EPR outcome.

The EPR Outcome in the Nonlocal Mindset

It's not as simple as rain falling from the sky.
—BACH & BELARDO

Truth only can be stated, not discussed. You only can decide on what truth is, if you know truth, so just state it.
—BACH & BELARDO

In the nonlocal mindset, A and B are not separated.

"Knowledge" is infinite or, in a local sense, omnipresent and does not depend on an entity that knows. A already has unlimited infinite information and knowledge and does not have to act upon arriving information. A does not act at all. It is the limited observer who selects one of the infinite realities that are there (not only at A, but everywhere). The concept here is that of an unlimited nonlocal mind where A does not act and where there is no change in A. It is up to an individual's information processing capability to access knowledge.

Because A is infinite, A knows all infinite possibilities and is not restricted to one particular state of being or outcome. An act of observation that commonly is thought to have an effect or to cause a disturbance does indeed change nothing; the observer picks one of the infinite possibilities that then becomes the one "observed reality"—but only for the observer.

A in a nonlocal unobserved state "knows" all the possibilities before anything is happening or will happen in a local reality. In a deeper sense, A and B are not, were not, and will never be separated. Looking at things this way suggests that knowledge is not created out of nothing.

Because of the inherent inseparability of A and B, there is no need for a signal or information to be transmitted.

Now the crux is to explain a nonlocal reality using local terminology:

We see A and B as separated. But the underlying reality remains nonlocal, meaning inseparable, which first of all cumulates into the realization that we have to consider a "local/nonlocal duality of reality." Nonlocal reality, in an incomprehensible sense, consists of an infinite number of realties from which an observer realizes one.

Technically speaking, as long as no observer is involved in the EPR experiment, the nonlocal reality (an infinite number of local realities) remains fully intact, and the probability wave function thus remains undivided in the half boxes at A and B, even though we see them separated. The half boxes A and B are separated in a local reality, but the unobserved inside of A and B remains nonlocal and undivided (expressed in a quantum mechanics equation—but only for an outside observer, not for an inside observer). That means there exists inside the two half boxes only one mathematically abstract nonlocal reality that consists of an infinite number of local realities. When an observer is looking into B, one of the infinite numbers of realities is chosen, and it seems that in turn the nonlocal reality collapses from infinite to one.

Again, information here is not physical. It is an intangible infinite unlimited reality like nothing we know, and the total lack of language is a challenge to give concrete meaning to any word. A particle, for example, does not have a finite number of predetermined local properties such as form, mass, expansion, etc., but it can take any finite form that is created by the limited intellectual capability of an observer and is not dependent on a predetermined set of classical forces either.

Knowledge here is the "selection of one reality" out of "an infinite number of realities." Knowledge is not created out of nothing. Whatever reality is chosen, it already existed.

Considering that progress in our way of thinking is desirable, it is imperative in the beginning to distinguish the two incompatible local and nonlocal mindsets and to develop nonlocal thinking and language in order to understand and improve life.

Beyond Modern Science: There Are Two Perspectives—You and Not You

This kind of overall way of thinking is not only a fertile source of new theoretical ideas: it is needed for the human mind to function in a generally harmonious way, which could in turn help to make possible an orderly and stable society.

—DAVID BOHM

It is anybody's task today to build a just society from the grassroots of humanity. Be a grass root of humanity!

—BACH & BELARDO

The issue that has been resolved by the EPR experiment goes beyond a mere technical matter between contending theories of subatomic reality. You cannot understand something by looking at its pieces. The significance of the EPR experiment lies in our conception of the objective universe and the nature of reality; thus, its results point far beyond the debate on quantum mechanics, Einstein's alleged locality argument, or the separability of objective reality. The outcome of the EPR experiment is not self-explanatory. It merely rejects locality without any indication as to what this means; local theories are insufficient to explain nonlocality and instant-action-at-a-distance. This creates the need to step outside of the framework in which the discussion has taken place and proceed beyond locality and the spacetime-dependent views of modern science. A fresh look at the situation from a new perspective might enable us to find phenomena in the EPR experiment that hitherto have not been recognized.

The remainder of the book focuses on interpreting the results of the EPR experiment and the phenomena of nonlocality and instant-action-at-a-distance. In local theories, instant-action-at-a-distance requires instant transmission of information, and from this it follows that information is the essential underlying principle of reality. Because there is a lack of relevant theories, it is necessary to articulate a nonlocal theory, independent of spacetime, that can coherently explain phenomena from the viewpoint of an information-based reality.

The Einstein-Podolsky-Rosen (EPR) Experiment: Turning Guns and Bullets into Roses

It's not that I'm so smart; it's just that I stay with problems longer.

—ALBERT EINSTEIN

Think long and hard—very long and very hard—and you can turn a bullet into a rose. They are the same—it is your choice.

—BACH & BELARDO

The following chapter is based on Paul Davies's description of the EPR experiment[19]. It details instant influence on a wave function at a substantial distance and provides a glimpse of what is meant by nonlocality and instant-action-at-a-distance.

In Davies's account, the EPR experiment consists of three ingredients: one "particle," a dividable "box," and a "probability wave function." Let us say the particle is a bullet. As long as we can see the bullet, we know the bullet's exact location in the box. If we close the box, we cannot look inside, but we assume the bullet becomes weightless and floats; we cannot know exactly where the bullet is. The only thing we can do is assume a "probability" as to where the bullet could be at any given moment. Nevertheless, we can assume a 100 percent probability (quantum wave or probability wave) that we will find the bullet in the box if we opened it. If we divide the box into two halves, A and B, there is a 50 percent probability that we will find the bullet in A and a 50 percent probability that we will find it in B. However, it is important to note that dividing the box into half boxes A and B gives us two 50 percent probabilities, but there is still only one probability wave function, which remains undivided. If we take A and B to the opposite sides of the universe, the probability wave remains undivided. If we open one box, the probability will become certainty, and the wave function will disappear. The question is, if we open B and find the bullet, how long will it take for the probability wave at A to learn about the observation that we have found the bullet at B and disappear (collapse)?

To illustrate, in Picture 3, a particle (1) is placed inside a box (2). The sealed box (3a) represents a closed system, meaning no communication with the outside is possible. Because the exact location of the particle cannot be known, it could be anywhere in the box with equal probability. This probability is expressed as a wave function (3b) that uniformly spreads throughout the box, representing the probability that the particle will be or will emerge at a certain location at a certain time. The box (4) is then divided into two half boxes (A and B). Apparently the particle is either in A or B now. A and B are then separated in space and time and brought to opposite sides of the universe. The probability wave in (5) remains inseparable and undivided, i.e., nonlocal, representing the probability that the particle is either in A or B and, unless observed, the undivided wave remains intact.

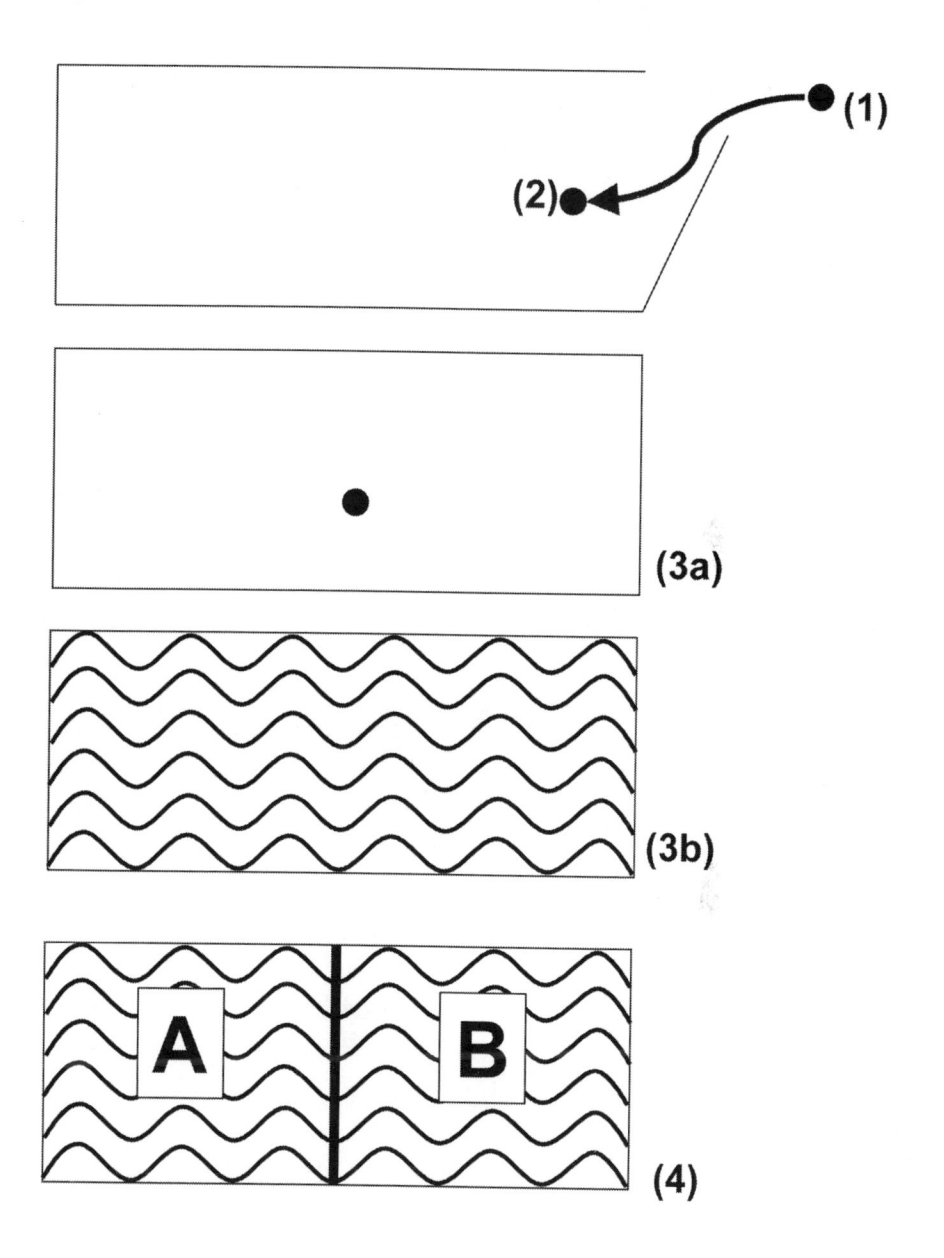

The Einstein-Podolsky-Rosen (EPR) Experiment: Turning Guns and Bullets into Roses

Picture 3: One particle (1) is placed in a box (2). When closed (3a), its associated probability wave is spread uniformly throughout the box (3b). A divider separates the box into two isolated boxes, A and B, and it is unknown whether the particle is in A or B (4).

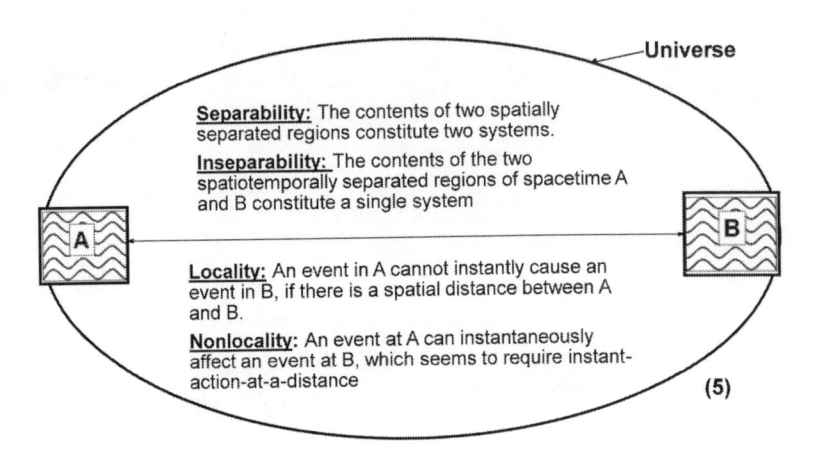

```
                    ┌─────────────────── Universe
        ╭───────────────────────────────────╮
        │   Separability: The contents of two spatially   │
        │   separated regions constitute two systems.      │
   ┌──┐ │   Inseparability: The contents of the two        │ ┌──┐
   │A │─│   spatiotemporally separated regions of spacetime A│─│B │
   └──┘ │   and B constitute a single system               │ └──┘
        │                                                   │
        │   Locality: An event in A cannot instantly cause an│
        │   event in B, if there is a spatial distance between A│
        │   and B.                                          │
        │   Nonlocality: An event at A can instantaneously  │
        │   affect an event at B, which seems to require instant-│
        │   action-at-a-distance                       (5)  │
        ╰───────────────────────────────────╯
```

Picture 4: A and B are brought to opposite sides of the universe. The single probability wave remains undivided (5).

In the separability argument, the contents of two spatially separated regions constitute two isolated systems if the distance between the two systems is big enough that the two systems cannot influence each other directly (proximity argument). The distance between the two systems consists of point-regions (smallest length of space) that have some concrete space-expansion. Crossing the space-expansion of each point-region inadvertently requires time. From this follows the locality argument: an event in A cannot instantly cause an event in B if there is a spatial distance between A and B.

Now, performing the EPR experiment, an act of observation will determine whether the particle is in B. According to the rules of quantum mechanics and contrary to the separability and locality arguments, at that instant, the wave abruptly disappears from A (see Picture 4), even though there seems to be a spatial distance to B at the other side of the universe.

The Einstein-Podolsky-Rosen (EPR) Experiment: Turning Guns and Bullets into Roses

To settle the matter of whether the wave collapses in A instantly or after some time depends on Bell's inequalities. If Bell's inequalities are not violated, no information has been instantaneously transmitted, and the wave in A would remain intact until a signal from B arrives at A. This would corroborate Einstein's locality argument and the separability of reality, and it would demonstrate that local events and things with local properties have an independent existence and do not influence each other. If the inequalities are violated, A would have been influenced by or have known about an observation occurring at B and would vanish accordingly. This would corroborate nonlocality and the inseparability of reality and demonstrate that information about local events is instantaneously known throughout the universe.

With regard to locality, the essential assumption promoted by the EPR experiment is that when an observation occurs at one of the locations (A or B), the other location, or anything in it, cannot have instant knowledge of this event, because information cannot be transferred faster than the speed of light. With regard to nonlocality, it is possible that as soon as an observation occurs at one box the properties of the other box change instantly. As illustrated in Picture 5, the EPR experiment shows that the inequalities are violated and thus seems to contradict Einstein's principle of locality and the notion of local realities. Therefore, changes induced in one physical system can have an immediate effect on another system separated by time and space.

The General Theory of Information

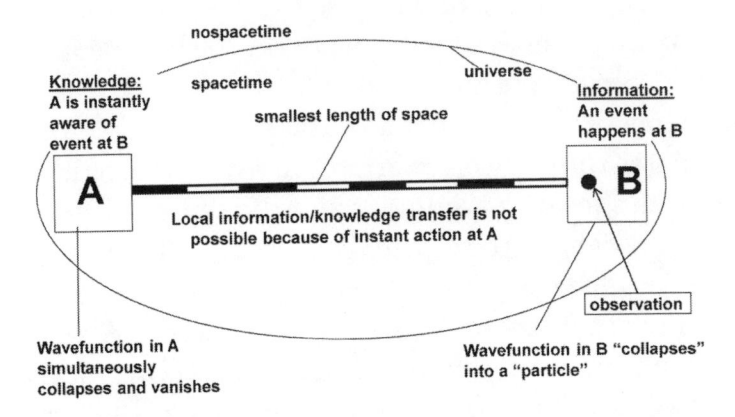

Diagram labels:
- nospacetime
- spacetime
- universe
- smallest length of space
- **Knowledge:** A is instantly aware of event at B
- **Information:** An event happens at B
- A — Local information/knowledge transfer is not possible because of instant action at A — B
- observation
- Wavefunction in A simultaneously collapses and vanishes
- Wavefunction in B "collapses" into a "particle"

When and how does A know that B was observed?
- know when: before, at the moment when and after an observation occurred at B
- how: nonlocal information of the past, present and future is omnipresent

Picture 5: Instant collapse of the probability wave, because the EPR experiment violates Bell's inequalities. Information about the event at B is instantaneously known at A.

The EPR result shows a violation of Bell's inequalities.

The result of violating Bell's inequalities means that the unobserved wave in A instantaneously knows when the act of observation occurs in B, which, in turn, would result in instant-action-at-a-distance. Yet, instant-action-at-a-distance cannot be explained simply by assuming that information about an event at B can travel through spacetime infinitely fast in violation of the special theory of relativity. This would also violate Einstein's locality argument, which posits that if there is a "distance," there must be a time lapse to bridge that distance.

Einstein Is Right, Despite the Error in His Thinking

In the strict formulation of the law of causality—if we know the present, we can calculate the future—it is not the conclusion that is wrong but the premise.

—WERNER HEISENBERG

If there is a distance, there is time needed to cross the distance. If there is no time, there is no distance. Easy—just state the obvious and think later.

—BACH & BELARDO

So why is Einstein correct despite the EPR experiment's results? There *is* no distance! It is the word "if" that makes the prerequisite "distance" irrelevant (not applicable). In this sense, Einstein's argument remains valid, because if there is a distance, it would take time to bridge it. The fact that no time is measured indicates that there is no distance. The special theory of relativity offers a partial solution, given that, under certain circumstances, time and space equal zero. Space and time vanish, and any distance disappears. Accordingly, the nature or reality that exists at the speed of light has something to do with a spacetime-independent reality, which causes the EPR experiment's result and which exists next to or beside a spacetime-dependent reality. It might be noted that Einstein did not comment on the nature of reality at the speed of light. Arguing that photons and tachyons have no mass at or beyond the speed of light is stating the obvious and does not explain anything.[18]

How else might one describe the EPR experiment? We believe that the EPR can be construed differently if we reorient our perspective and think about how the experiment applies to information in the sense that any kind of information is omnipresent and undivided throughout space and time and extends beyond the local universe. We believe, therefore, that information can exist beyond the local universe even if space and time do not.

Space-Time Dependent Interpretation: The Local Mindset of Normal Science

Everything we call real is made of things that cannot be regarded as real.

—NIELS BOHR

Normal science is not wrong, just very restricted, as it needs to follow the logics of separability.

—BACH & BELARDO

According to normal science in the Kuhnian sense, instant-action-at-a-distance would require information to travel faster than the speed of light. Instant-action-at-a-distance surprisingly remains a local theory despite the logical requirement of infinitely fast proliferation. A paradox (Bell calls it muddle) like this encourages scientists to focus on prediction and control rather than on understanding and explanation.

> Paradoxes are accepted as normal, but they severely affect genuine progress in science by shutting down further thinking and reflection needed to develop a variety of reasonable resolutions and solutions that might lead to progress in science.

The local approach is constrained by the assumption that spacetime is perpetual (nonvanishing, always applies) and that nothing can exist outside of observable spacetime. In this way, scientific interest limits itself to what can be observed. This ultimately diminishes progress in science over time and is the reason why science progresses in revolutionary leaps.[10] Quantum mechanics and Bell's Theorem, however, prove "inseparability" even though it cannot be observed.

In local understanding, the transfer of information over a distance between two points separated in space must occur using time; this excludes instant connectivity. An infinitely fast transfer of information cannot be satisfactorily explained by local theories. In local spacetime theories, nonlocality would mean that all points in space are somehow connected, regardless of their spatiotemporal distance from one another.

> In order to explain this violation of the special theory of relativity, information must have nonlocal properties that are independent of spacetime.

Space-Time Dependent Interpretation: The Local Mindset of Normal Science

Ultimately, this means that information exists beyond space-time and, in a sense, is omnipresent. However, the omnipresence of information is still a local concept, meaning that information is somehow spread throughout space and time and that each point in space includes all information equally. Here again we are constrained by the limitations of language.

> In the EPR experiment, as soon as the box is closed there is no particle in either box, and there is no probability wave.

As soon as the system is unobserved, information is nonlocal and exists everywhere simultaneously, not merely as a wave function. If there is no need for a particle to be at a certain location, there is no need for a wave function.

SPACE-TIME INDEPENDENT INTERPRETATION: THE NONLOCAL MINDSET

What we observe is not nature itself but nature exposed to our method of questioning. Our scientific work in physics consists in asking questions about nature in the language that we possess and trying to get an answer from experiments by the means that are at our disposal.

—WERNER HEISENBERG

If you don't know what nature is, you might ask questions in a way that leads to the wrong interpretation. More precisely: you ask a question, you do an experiment, and you obtain a result that is the answer to a different question, but when you base your answer on that question, you arrive at a wrong conclusion. For example, you ask the double slit experiment if something is a particle or a wave, and the experiment tells you that it is neither. The question, not the experiment, puts the notion of particle and wave into the experiment.

—BACH & BELARDO

The EPR arrangement of two distant systems (A and B) illustrates the true nature of nonlocality in the sense that instant-action-at-a-distance is a misleading concept and transcends the notion of signals traveling faster than light.

> In fact, the simultaneous collapse of the wave in A and B demonstrates the unthinkable, namely that instant-action-at-a-distance, in a sense, does not require any signal to be sent.

The arrangement accurately depicts inseparability. The probability wave in A and B is not separated. It remains undivided even though A and B are separated in space and time on opposite sides of the universe. This phenomenon is counterintuitive, even though the mathematical formulation describes the wave as undivided. The same happens in the two-particle-system in Bell's Theorem, where the measurement at one particle instantly adjusts properties (e.g., polarization, or spin) at a distant counterpart particle. Intuitively, one would expect that knowledge of a measurement at the observed particle to have somehow been transmitted to the unobserved particle. Yet, because of the omnipresence of information, this is not the case.

The next chapter illustrates the nature of omnipresent information, showing that information about the future is already known in a nonlocal, unobserved state. Subsequently, nonlocal theories do not require the signaling or transmission of information. One consequence of this is that one can choose from an infinite number of possibilities at any time. Local reality is thus not pre-determined, and there are an infinite number of possibilities from which to choose at the crossroads of each and every decision.

In summary, based on the notions of instant-action-at-a-distance, instant connectivity, omnipresence of information, and nonlocality/inseparability of reality listed in Table 1, one arrives at the new phenomenon of "nonlocal information," which is the ubiquitous, inseparable information that constitutes reality.

Phenomenon	Definitions
Nonlocality	• Inseparability • Inseparability of reality
Instant action-at-a-distance	• Replaced with omnipresence of information in a local reality
Omnipresence of information	• Nonlocal information

Table 1: List of phenomena essential for formulating our General Theory of Information.

EPISODE 3

Time is what prevents everything from happening at once.
—JOHN ARCHIBALD WHEELER

At the speed of light, time equals zero. Space equals zero. Nothing changes in the nonlocal reality. Everything happens in an infinite number of parallel local realities.
—BACH & BELARDO

Riding the Cannonball of Light: Formation of an Information Singularity at the Speed of Light

If you thought that science was certain—well, that is just an error on your part.

—RICHARD P. FEYNMAN

Nonlocality, whatever it is, is the only certainty—even if we don't understand it.

—BACH & BELARDO

The relativity of time and space offers new insights into the concepts of nonlocality and nonlocal information. The language used in this chapter is a combination of local and nonlocal wording and logic, because a local observation may be looked at and defined using both local and nonlocal logic. For example, when the mass of an object becomes infinite, the object itself becomes infinite, and this violates local logic. Using nonlocal logic and replacing the local term "mass" with "nonlocal information," it becomes more accessible that the information content of an object is infinite, which is exactly what the special theory of relativity correctly predicts, even though this cannot be understood using local logic. The special theory of relativity predicts exactly what was observed decades later in the EPR experiment.

The key observation and insight derived from the EPR experiment, however, is that "inseparability" is de rigueur, which means that an object not only consists of an infinite amount of information at the speed of light but also below the speed of light. The true nonlocal nature of an object never changes; it always consists of an infinite amount of inseparable information or other properties, such as mass and energy. The differences in observation of an object with less than infinite mass are the result of relativity. Things look different because of the relativity of observation, which creates the finite and changing nature of things. Only under certain circumstances, or at specific vantage points, would the observer be able to see the true nature of things. In the case of the special theory of relativity, this would be at the speed of light. The observation of the true nature of an object, however, requires the simultaneous processing of an infinite amount of information, of which a local observer is not capable. It is only through indirect scientific and common sense inferences and reflections free of preconceptions about relativity and inseparability (e.g., "infinite mass at the speed of light" and "instant-action-at-a-distance") that one can derive absolute knowledge and truth. Our ability to reach this goal is thwarted by the coerced uniformity of thinking in science, which allows only one system of logic (separability, fragmentation) and one perspective (separation from nature, creation out of nothing). Words and definitions are therefore tentative and need to be redefined in each new context.

Phenomenon	Definitions
Nonlocality	• Inseparability • Inseparability of reality
Nonlocal information	• Omnipresent in a local reality • Exists beyond the speed of light in a nonlocal reality
Time	• Equals zero at the speed of light
Space	• Equals zero at the speed of light

Table 2: Phenomena and definitions used in this chapter.

By considering the phenomena in Table 2, one might be able to determine the behavior of time, space, and information at or beyond the speed of light as shown in Picture 6. The illustration in Picture 6 is based on the scientific observation or convention that time equals zero at the speed of light.[20] The questions are: what happens to time, space, and information when they reach the speed of light and what can be known about what exists *beyond* the speed of light?

From the equation $E=mc^2$, it follows that the mass of an object becomes infinite when it reaches the speed of light; the corollary of which is that no object can actually reach the speed of light. Therefore, because no observation can be made beyond the speed of light, normal science cannot explain reality; it remains scientifically impossible to define a local reality. In this sense, the speed of light is a barrier beyond which reality is inexpressible and beyond the paradigmatic scope of normal science. But this conclusion is refuted by the fact that photons obviously do travel at the speed of light and can be observed doing so. There is, however, no theory available that can satisfactorily account for this observation.

What Exists at the Speed of Light

Science no longer is in the position of observer of nature, but rather recognizes itself as part of the interplay between man and nature. The scientific method ... changes and transforms its object: the procedure can no longer keep its distance from the object.
—WERNER HEISENBERG

Man is Nature.
—BACH & BELARDO

In order to exist at the speed of light, photons (corpuscular light in the form of a particle or object) must have in a local mindset "no mass," or in a nonlocal mindset "exist outside of spacetime" (with nonlocal properties) but must also be present in both the local and nonlocal reality. A photon's information is present at a *certain* point in space and time and at *every* point in space and time. This implies that the omnipresence of information is not disturbed by the act of observation. In short, when you observe something you don't disturb it because there is nothing to disturb. However, your observation creates something by limiting nonlocal information.

The General Theory of Information

Collapse of Time, Space, and Information at the Speed of Light

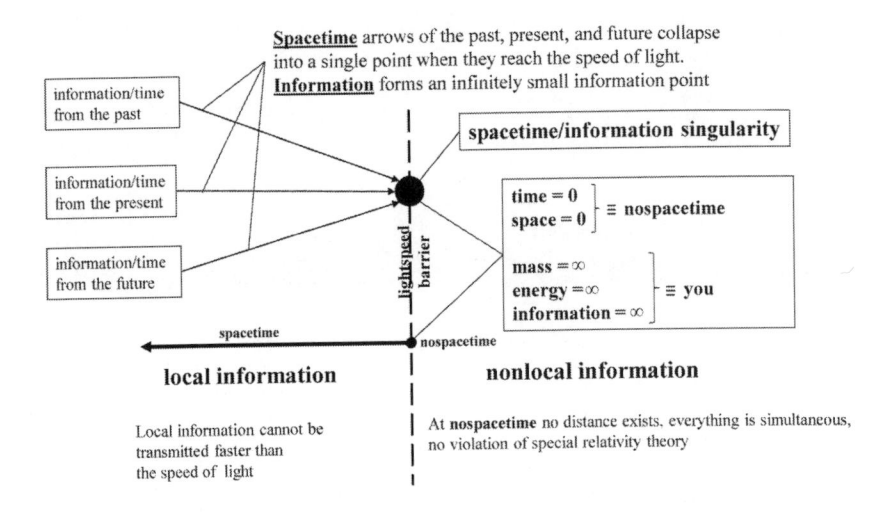

Spacetime arrows of the past, present, and future collapse into a single point when they reach the speed of light. **Information** forms an infinitely small information point.

- information/time from the past
- information/time from the present
- information/time from the future

spacetime/information singularity

$\left.\begin{array}{l}\text{time} = 0 \\ \text{space} = 0\end{array}\right\} \equiv \text{nospacetime}$

$\left.\begin{array}{l}\text{mass} = \infty \\ \text{energy} = \infty \\ \text{information} = \infty\end{array}\right\} \equiv \text{you}$

lightspeed barrier

spacetime ← | → nospacetime

local information | **nonlocal information**

Local information cannot be transmitted faster than the speed of light

At **nospacetime** no distance exists, everything is simultaneous, no violation of special relativity theory

Picture 6: The behavior of space, time, and information at the speed of light. Reality beyond the speed of light is defined as nospacetime and information singularity.

When it reaches the speed of light, an object's mass becomes infinite. If the mass of an object is infinite, its information content at the speed of light must also be infinite. Both properties behave in a similar fashion. Information content increases as the object moves. Space and time, however, behave differently.

> The faster the object travels the more space contracts and the slower time goes. This leads to the paradoxical outcome in which an object increases in mass while shrinking in size!

This underscores the fact that the words object, mass, and size become meaningless when they are taken outside of the experimental context in which they have been created and defined. Such terms have neither "independent properties" nor "independent values," or "independent existence" as Einstein argued. In sum, the faster an object travels, the more its mass increases and its size decreases while space continuously contracts until the object vanishes at the speed of light.

A solution to these contradictions lies in the inseparability requirement of reality in the EPR experiment, i.e., information about the past, present, and future is inseparable and constitutes an indivisible local spacetime continuum. In the indivisible continuum, the hypothetical time vectors of past, present, and future remain inseparable and affect each other. Suppose an object accelerates to the speed of light. The special theory of relativity predicts that the greater the speed the more time slows and space shrinks, thus the time vectors in Picture 6 move toward zero, and time and space vanish. Information, however, does not vanish but collapses into a reality beyond spacetime and forms an inseparable information singularity. The illustration in Picture 6 shows that when the speed of light is reached, spacetime collapses into its converse, nospacetime, and forms an "inseparable information singularity." Thus, from an information perspective, reality becomes nonlocal beyond the speed of light and forms a nonlocal reality, which might be viewed as a "singularity of information in nospacetime."

The behavior of information differs from that of time and space, because information can travel faster than the speed of light (even in the concept of instant-action-at-a-distance). Accordingly, an information-based reality can exist beyond the speed of light. Information from the past, present, and future does not vanish at the speed of light but collapses into an information-singularity of nonlocal information. These observations are summarized as follows in Table 3.

Phenomenon	Definitions
Nonlocality	• Nospacetime • Spacetime singularity
Nonlocal information	• Information singularity • Infinite amount of inseparable information
Time	• Vanishes at the speed of light
Space	• Vanishes at the speed of light
Information	• Remains at the speed of light • Information of the past, present and future collapses (becomes infinite) at the speed of light

Table 3: Phenomena and definitions derived from this chapter.

Based on the concepts of nonlocality, nonlocal information, information singularity, and nospacetime (spacetime singularity), it is now possible to formulate our General Theory of Information. This is the subject of the following chapter.

EPISODE 4

Jordan declared, with emphasis, that observations not only disturb what has to be measured, they produce it. In a measurement of position, for example, as performed with the gamma ray microscope, 'the electron is forced to a decision. We compel it to assume a definite position; previously, it was, in general, neither here nor there; it had not yet made its decision for a definite position .... If by another experiment the velocity of the electron is being measured, this means: the electron is compelled to decide itself for some exactly defined value of the velocity ... we ourselves produce the results of measurement.'
—J. S. BELL

... in the experiments about atomic events, we have to do with things and facts, with phenomena that are just as real as any phenomena in daily life. But the atoms or the elementary particles are not as real; they form a world of potentialities or possibilities rather than one of things or facts.
—WERNER HEISENBERG

Observe nature, whatever it is, from the perspective/reference frame of light. There is no reasonable definition of nature.
—BACH & BELARDO

Two Worlds That are One: The General Theory of Information

Nature is made in such a way as to be able to be understood. Or perhaps I should put it—more correctly—the other way around, and say that we are made in such a way as to be able to understand Nature.

—WERNER HEISENBERG

Why believe that nature is something else "out there" and that we are not part of nature? All evidence since Thomas Young in 1800 shows the opposite.

—BACH & BELARDO

The EPR experiment provides experimental evidence that underlying instant connectivity is an information-related phenomenon. The properties of this phenomenon have escaped the attention of normal science[10], because the standard theoretical frameworks, as well as the dominant philosophical and methodological perspectives, are inattentive to certain phenomena, particularly information-related ones[18]. For this reason, we have developed a novel theory of information, which focuses on the nature of information, the characteristics of which include non-localizable, omnipresent properties independent of spacetime. The phenomenon of instant transmission of information in the EPR experiment is explained through the omnipresence of spacetime-independent nonlocal information.

The EPR experiment shows that information about a local event is known instantaneously throughout the universe. From the perspective of local spacetime-based theories, this requires that information travels infinitely fast and that objects cannot have an independent existence, even though they appear to be localizable in the perception of a local observer with limited information processing capability. The physical localizability of objects does not imply that they exist beyond observation; even when localized, when they seem to have an independent existence, it is only relative to an observer's position. The underlying instant connectivity is invisible and does not constitute a physical process that can be described fully by a theory based on local spacetime. A local theory assigns properties to enduring physical objects at each point on their trajectory through space[9]. Traveling through spacetime requires a time lapse; therefore, the instant transmission of information in the EPR experiment cannot be explained by theories that depend on local spacetime. In the information perspective of a nonlocal nospacetime theory independent of observation, the phenomena of nonlocality and the omnipresent existence of information in the EPR experiment provide evidence of an inseparable information-based reality that has the property of instant connectivity.

A complete theory must be able to define both local and nonlocal reality, nonlocal processes and phenomena, and the locality and localizability of observable objects and their properties.

The General Theory of Information: Two Stories to Tell

Any reasonable idea is worthless.
— ALBERT EINSTEIN

A nonlocal mindset is beyond reasonable.
— BACH & BELARDO

Based on the phenomena of nonlocality and nonlocal information derived from the EPR experiment, the General Theory of Information defines the fundamental nature of reality as consisting of inseparable nonlocal information, where everything about anything is instantaneously known everywhere. Nonlocal information is defined as an "infinite amount of inseparable information." Because of the inseparability of reality, perceiving this information requires the capability to process an infinite amount of information simultaneously. Localizability is a finite information process that limits information and causes the creation of local realities.

Definition of the General Theory of Information: Don't Worry—You Can Use It!

If we knew what it was we were doing, it would not be called research, would it?
—ALBERT EINSTEIN

New thinking for a new world without hunger and poverty.
—BACH & BELARDO

If we are to take local and nonlocal phenomena into account, a complete theory must include and satisfy two criteria: an assumption regarding the existence of nonlocal reality and an explanation for the creation of local realities. The General Theory of Information addresses these issues in the following statements:

> 1. Nonlocal information constitutes a nonlocal reality consisting of an "infinite amount of inseparable information."
> 2. Finite information processes limit nonlocal information and create local realities.

This definition is not meant to be exclusive or definitive. Indeed, the word information, as in the phrase "infinite amount of inseparable information," can be replaced with many other synonyms and yield the same result. An infinite amount of inseparable mass, for example, would lead to the same conclusion. The result is still "everything." What can be observed is determined by the viewer's relative point of view and capacity to process information, not by the nature of an elusive object or phenomenon. In Table 4, the definitions of the various nonlocal concepts are recapitulated.

Phenomenon	Definitions
Nonlocality	**Nonlocality** is defined as the inseparability of reality. At the speed of light it forms an information singularity in nospacetime.
Nonlocal Information	**Nonlocal information** is defined as an "infinite amount of inseparable information." It includes the phenomenon of the inseparable omnipresence of information.
Localizability	**Localizability** is defined as the act of limiting the infinite amount of inseparable information. An observer with limited information processing capability creates a local, observer-dependent reality.
Local-Nonlocal Dualism	**Local-Nonlocal Dualism** is defined as the complementary existence of local and nonlocal reality.

Table 4: Phenomena and definitions supporting the General Theory of Information.

Definition of the General Theory of Information: Don't Worry—You Can Use It!

The General Theory of Information posits the existence of two realities, one local and one nonlocal. This dualism needs to be separated, however, because the logic associated with each is different and this creates confusion.

Table 5 lists a few implications of the local/nonlocal dualism of reality. We attempt to demonstrate how the same terms can be defined differently using local as opposed to nonlocal logic, and there may be overlap. The statement "nothing can travel at the speed of light, except light itself" exemplifies this dualism, because the phrase "except light itself" does not fit the paradigmatic framework of normal science and, therefore, is not subject to scientific inquiry.[18]

Fundamental consequences of the local/nonlocal dualism of reality:

Concept	Local Reality	Nonlocal Reality
Creation	Creation out of nothing	Neither creation nor evolution, nonlocal information
Evolution	Random evolution	Evolution in planned leaps and guided incremental steps
Reality	Finite separable reality	Infinite inseparable reality
Double Slit Experiment	Observed electron in a local reality produces a Gauss distribution curve	Unobserved electron in a nonlocal reality produces an Interference pattern
EPR	Instant action at a distance	No distance
Relativity	Local reality below the speed of light	Nonlocal reality beyond the speed of light
Zeno's Paradox	A real "object" jumps from one point in space to another point in space	An hypothetical "object" jumps between the local and nonlocal reality to create movement Quantum Information Dynamics (QID)

Quantum computer	Infinite parallel information processing	Information processing in nospacetime requires no time
Human Genome	Contains limited information	Contains nonlocal information with the ability to fold and unfold information to form an infinite variety of structures. The origin is a repeating fractal or origami element
Society	Individuals competing with each other creating fragmentation, war, and poverty	Cooperation and symbiotic creativity
Origin	Nothing—Beginning/Creation—Evolution—End/Nothing Before the beginning was nothing. Something is created out of nothing that evolves and ends.	Infinity—Has always existed—no beginning or end • No creation • Planned evolution; unfolding information in an infinite number of forms

Table 5: Fundamental differences between the local and nonlocal mindset.

It should be noted that the limitations of language tend to obscure the differences between the fundamental logic of local and nonlocal reality. The reader is advised to reflect upon the content of the words rather than the words themselves, which are malleable and time specific.

CAN WE UNDERSTAND THE PHRASE "INFINITE AMOUNT OF INSEPARABLE INFORMATION"?

We have heard it so many times: there is nothing new in this book. This and that have already been said here and there. Yes, we can cite thousands of books and texts of the last thousand years since writing has emerged that say exactly what is in this book. But, after thousands of years, what have we accomplished? Nothing— no progress for humanity. We still kill each other and use technology and knowledge to do so more effectively. We still wage war against whatever we think nature is and are ever more efficient to make this world uninhabitable for us—nature does not care. Nature can blow up entire universes and start from scratch. We have to stop talking, writing, discussing, fighting—and do the right thing: working together as one for the prosperity and happiness of the One.
—BACH & BELARDO

Why not use science to finally do the right thing?
—BACH & BELARDO

Is the phrase "infinite amount of inseparable information" the ultimate solution? Probably not. It is just a string of words to make sense of. We can somewhat understand the meaning of each individual word—"infinite," "inseparable," and "information"—but as soon as we arrange these words into a group and start thinking about the implications of this phrase, namely that each of us consists of an "infinite amount of inseparable information," this becomes speakable but unthinkable. We might want to say that it does not make sense, but doing so would shut down thinking and prevent progress. There are an infinite number of perspectives or point of views to look at this string of words.

We also have to ask what good is knowledge without understanding. At least we know something and should want to progress with our understanding in terms of spending our efforts on a new thinking and a new mindset.

Summary:

1. We can know of an "infinite amount of inseparable information."
2. We cannot understand an "infinite amount of inseparable information."
3. We need to develop a new thinking and a new nonlocal mindset.
4. We need to keep the local and nonlocal mindset strictly separated.
5. There is no new thinking and no nonlocal mindset as yet, only together can we advance and develop it.

EPISODE FIVE

Suppose we were able to share meanings freely without a compulsive urge to impose our view or conform to those of others and without distortion and self-deception. Would this not constitute a real revolution in culture?

—DAVID BOHM

A new world in the making.

—BACH & BELARDO

We are good at using science and technology to efficiently and effectively turn our natural resources into piles of trash. We have to use science and technology differently.

—BACH & BELARDO

What Now? Everything at Your Fingertips: Let's Work Together

We can't solve problems by using the same kind of thinking we used when we created them.

—ALBERT EINSTEIN

Even if we would develop absolute knowledge, we might not be able to understand it.

—BACH & BELARDO

The General Theory of Information depicts the creation of reality as an information process through the limitation of nonlocal information (an infinite amount of inseparable information) by an observer with a limited information-processing capability. Reality is the product of the limited information-processing capability of a finite mind. The General Theory of Information supports the position that there is an objective reality that exists beyond the human mind in the form of nonlocal information and from which an individual is never separated. The challenge of inseparability to common sense is that there is no reality "out there" to be discovered. Any object, whether animate or inanimate, consists of the same undivided and undividable information. The scientific method has its limits and is partially responsible for the fragmentation of being and the poor state of the world as discussed by Bohm.[21]

The implications of the General Theory of Information are far-reaching in allowing researchers and thinkers to identify and articulate new phenomena that previously have not been identified because they lie outside the paradigmatic realm of normal science.[10]

Because everything is made of the same undivided and undividable information, the General Theory of Information strongly suggests that we belong together undividably (sorry no choice here) and that a new "constructive and collaborative thinking" that is inclusive and creates a feeling of belonging will be the agent of change for a permanently better world.

The "new constructive science" now moves beyond fragmentation of thinking and practice, where each of us is a small piece in a big machine, and towards mutual care and collaboration. The goal is to build a global platform that provides the knowledge and technology to the people living in poverty, enabling them to produce the basics to ensure a fruitful and fulfilled creative life in their local culture that at the same time lets them participate in a global system of creating mutual prosperity in a healthy and sustainable way. Eliminating the struggle of creating the basics for living will deliver freedom to all, open up creativity, and foster local cultural diversity and the development of a mutual inclusive participatory platform.

History shows that the reliance on technology, knowledge, and competition has repeatedly caused widespread calamities and brought the world to the brink of destruction again and again. With the right creative use of technology, knowledge, and collaboration, these cycles of destruction can be broken. This civilization has its share of challenges, but with technology, knowledge, and a mutual *will*, we the *people* now have the chance, for the first time, to make a permanent turn for the better for all—not just for a few. By collectively taking life matters into hand and not letting others (external observers) dictate the basic settings of how this world works—to the advantage of the few, leaving the majority in poverty—we can use this moment to build the foundations of truth, trust, collaboration, and mutual and shared prosperity that allows *all people* to live a life of dignity. The premise is that it must include *all* people. In this sense, the General Theory of Information helps to bring about social improvements, as it closes the gap between life and nature, between environment and technology.

Ex Nihilo Nihil Fit:
Birth and the Mystery of Life

It is the plunge by which we gain a foothold at another shore of reality. On such plunges the scientist has to stake bit by bit his entire professional life.

—MICHAEL POLANYI

... anyone who fears being mistaken and for this reason studies a 'safe' or 'certain' scientific method, should never enter upon any scientific enquiry.

—KERLINGER

Not having a better answer does not compel one to believe—especially in the face of absolute knowledge. Not liking truth is not a scientific argument to believe in the opposite.

—BACH & BELARDO

Ex Nihilo Nihil Fit—Nothing Comes From Nothing

> *A fool thinks himself to be wise, but a wise man knows himself to be a fool.*
> —WILLIAM SHAKESPEARE

> *A genius is an idiot who knows that he is an idiot. And understands why!*
> —BACH & BELARDO

It is not like the General Theory of Information makes the unthinkable thinkable and the unspeakable speakable. It is more that we can start dealing with and looking constructively at unthinkable matters from an infinite number of new perspectives by using any kinds of paradigms that may serve us. We are going beyond the restriction of having to agree on one paradigm and one way of thinking.

In this context, and to continue with unthinkable thought, the widespread and commonly accepted notion that we are coming out of nothing is particularly challenging.

Any kind of approach to tackle what nothing *is* is hampered by the limitation of language, with its imprecise definition of words, because we cannot identify or determine a finite set of exact qualities or meaning that would define a word. The General Theory of Information actually advises that we need an infinite set of qualities and meanings to precisely define a word that is unthinkable and impracticable.

A beginning might be—if we assume that nothing comes from nothing—that we still need to define what nothing is and prove that nothing exists before we can propose a mechanism whereby something can be created from nothing.

In a more thorough reflection, we first have to admit that if we assume that nothing comes from nothing, there must be "a nothing," pure nothingness, and there cannot be something.

If we assume that something can be created out of nothing, everything can be created out of nothing; ergo nothing equals everything. The words nothing and everything have the same intrinsic indistinguishable meaning, or else it is not possible to ascribe meaning to the word nothing.

Now when we consider that only "everything" exists and "nothing" does not, neither creation nor evolution *per se* are possible or required and "everything" has the qualities of "inseparability" (EPR experiment) and "infinity" (at the speed of light)—no beginning and no end.

Without the inseparability argument of the EPR experiment, it would be conceivable that everything can be created out of nothing, whereby nothing marks the beginning, and things could emerge indefinitely on a time line. Without the relativity argument, time would be absolute, meaning that time does not shrink when moving faster, which would allow things to emerge out of nothing on a straight immutable time line. Both EPR and relativity are arguments against the possibility that something can be created out of nothing and against the existence of nothing itself.

Can we understand this? No. We do not have the intellectual capability to understand it. We can know it but not understand it; neither do we have the language to express it properly. The only thing we can do is to realize what follows from the EPR experiment, namely that time equals zero at the speed of light, without being able to truly understand and express this in words.

The General Theory of Information, in essence, conveys the solution that nothing can be added to infinity. Infinity is the start without a start and the notion of nothing is, for the lack of another word, unthinkable or, even if it was thinkable, not applicable.

BIRTH AND THE MYSTERY OF LIFE

Does not quantum theory again place "observers"... us ... at the center of the picture?... The general public could get the impression that the very existence of the cosmos depends on our being here to observe the observables. I do not know that this is wrong. I am inclined to hope that we are indeed that important.

—J. S. BELL

Everybody wants us to explain the unthinkable. If it would be explainable, it wouldn't be unthinkable, right?

—BACH & BELARDO

In one form or other, the basic aim of the scientific method is the discovery of truth, next to other functions such as the explanation of natural phenomena.[23] It is of great importance that the scientific method facilitated the discovery and experimental proof of absolute truth, but the scientific method has its limits, which lie in the fragmentation of thought and action, and it is equally important to understand this.

Scientific prediction and control lead to the observation that we are "born" and "die." We can predict with complete certainty that birth and death will happen and conclude from it that they are "real phenomena." But what are the phenomena we are observing? We see something and call it "birth," but observing and predicting birth does not create any knowledge about what birth actually is. We might ask what was before we came into existence. Do we really come out of nothing and go into nothing? The combination of two things creating a new thing is another form of creation out of nothing.

Does the General Theory of Information help resolve the conundrum of "creation out of nothing"? We argue that the information inherent in living beings exists before creation and remains after death. Birth and death can be observed and described but do not provide knowledge or understanding about any underlying truths. Formulas, theories, and statements do not suffice. What we require is the ability to think without constraints, free of preconceptions.

It might be helpful if we conceive of nonlocal information as a vast library containing all of the information there is about the past, present, and future; then we might imagine individual lives as the information contained in these books. The birth process would commence by unfolding the information contained in a specific book. Conception would trigger a process of limiting the nonlocal information contained in this volume and releasing it as local information via the human genome.

Unfolding Nonlocal Information via the Human Genome: Complexity and Simplicity

Science is about the honest realization how little we know.
—ALBERT EINSTEIN

Science delivers truth—likeable or not.
—BACH & BELARDO

Since the General Theory of Information argues that all information exists at all times everywhere, then it follows that the human genome consists of "an infinite amount of information," which can be folded and unfolded in limitless ways according to the dictates of nature. The sequence of three billion DNA base pairs, however, is not the human genome per se, but rather the apparatus through which nature rearranges and unfolds the information needed to build an organism. Through rearrangement, an infinite amount of information can be managed and expressed in an infinite variety of forms, using the human genome as the intermediary between local and nonlocal reality. As the Nobel Prize winning scientist Manfred Eigen argued, even a small gene consisting of one thousand base pairs has a vast number of possible combinations: 10^{605} combinations to be exact.[24, 25]

For some time, Eigen wondered how nature could physically produce and test 10^{605} combinations in spacetime to find the optimal genetic combination. According to his calculations, the entire biomass produced by nature in the history of our planet would produce approximately 10^{50} combinations of this one gene. Eigen noted that an event with a probability of 10^{50} is mathematically considered impossible to happen in all eternity. It seems highly unlikely that the random production of even a fraction of 10^{605} combinations could result in a practical or useful arrangement of matter producing a functional gene. Even if we would assume that matter creates life we do not really know what the terms "matter" and "life" signify and how a combination of inanimate matter can produce life just by combining itself.

From an information perspective, one might try to discern the "degree of information processing capacity of matter," in which matter is thought to process information by randomly arranging itself into various forms while creating new life. Yet there is no scientifically valid indication that matter has the independent capacity to do such a thing. In the case of 10^{605} combinations, matter would process information by randomly rearranging the one-thousand base pairs to produce, test, and select a functional gene, and even assuming self-organization and hypercycles,[26] it would not significantly reduce the number of 10^{605} to a practical level where matter can test a reasonable number of combinations.

In contrast to matter that processes information by combining its elements, we are "processing information by thinking," yet thinking has no scientific meaning, and the information properties and processes that constitute life must be evaluated indirectly through "constructive reasoning and reflection."

Life, according to the General Theory of Information, might be viewed as the local manifestation of information in which the "human genome serves as the information processor," translating nonlocal information into local components to form an organism. It is important to note that the process of forming an organism does not include the creation of the mind or of consciousness. Mind and consciousness have nonlocal properties that stay nonlocal in an organism but produce manifestation in a local reality through an intermediary device, such as the brain that translates nonlocal information into local manifestations. Evidence for this can be observed by disabling the speech center of the brain. Even if one's speech center is disabled, one can still keep counting mentally. This shows that the brain is the information processing apparatus that translates information from the nonlocal mind into local manifestations associated with moving, dreaming, or thinking. The information processing capability of the brain is limited, whereas the nonlocal mind is able to simultaneously process an infinite amount of information. The ability to process and express information is what distinguishes a particle from the human genome, both of which consist of an infinite amount of information.

Complexity and Simplicity

We are interested in truth; science is just a tool that delivers truth. Now we have to use truth and science to deliver the good for humanity—never has this been achieved before.
—BACH & BELARDO

We can do it ... period! Don't think further.
—BACH & BELARDO

The figure 10^{605} represents a dimension that cannot possibly be managed in a local reality by random or non-random arrangements of matter.[25] The exact arrangement of a gene consisting of one thousand base pairs would be unimaginably complex if one worked from even a fraction of those 10^{605} combinations. How would nature even begin to design the components of life? The possible combination of 10^{605} strongly indicates that nature utilizes simplistic rules to produce modular structures[27] and operators[28] with a high degree of elastic malleability that can be turned into very different functional units by applying minute structural or genetic changes. Examples of underlying simplicity can be found in the growth of fractals and in origami where one sheet of paper can be folded in unlimited ways to create endless forms. It has been argued that the human genome has an underlying fractal structure that repeats itself in a modular fashion,[29] and with repeated folding and unfolding,[30-35] information might be reversibly archived and revitalized. Modularity is thus an evolutionary trait that is used extensively by nature to cope with complexity. Modularity is the repeated use of simple elementary modules that evolve like a landscape out of a simple fractal or a repeated elementary fold in origami.

The human genome has both a nonlocal and local character; the former can be viewed as a Gestalt that only works in its nonlocal totality and cannot be understood by looking at its parts. Looking at base pairs and genes does not reveal the full nature of the human genome. As a Gestalt, the human genome acts according to a guiding principle and is not random. A Gestalt can, under purposeful evolutionary pressure, create the functional components for any kind of organism and entities of nature. The Gestalt of the human genome functions like an organism, expressing an interminable variability of forms and systems and capable of continual dynamic formability.

Finally, we use the preceding observations to answer two questions:

(1) Is there experimental evidence that might provide a sense of what can be considered absolute scientific knowledge?

Science has provided experimental proof for absolute knowledge, but it has been overlooked because it does not fit the paradigmatic framework of normal science.[10] Absolute knowledge might be expressed as: "infinite amount of inseparable information." In this manuscript, the EPR experiment and relativity were used to arrive at absolute knowledge. The same can be accomplished using the double slit experiment and other experimental findings.

(2) **Does absolute knowledge let us understand the absolute truth about ourselves?**

Truth is about understanding human nature. We believe the truth is that each human being consists of the same infinite amount of inseparable information, and it follows that we must engage in ethical behavior that reflects our true nature. Inseparability encompasses undividable responsibility: the choices that we make define the state of this world and have a direct impact on others and the entire universe.

Episode Six

A superior man is modest in his speech, but exceeds in his actions.
—CONFUCIUS

It is very easy to destroy. It is very difficult to build.
—BACH & BELARDO

If we have no peace, it is because we have forgotten that we belong to each other.
—MOTHER TERESA

There is no progress in social competition. True progress lies in genuine collaboration.
—BACH & BELARDO

WHO YOU ARE—WHAT YOU HAVE TO DO

Indeed, the attempt to live according to the notion that the fragments are really separate is, in essence, what has led to the growing series of extremely urgent crises that is confronting us today.

—DAVID BOHM

Humanity starts with each of us and our collaborative action that results in creating shared and mutual prosperity for all.

—BACH & BELARDO

The infinite resource of knowledge is freeing the world from hunger, poverty, and injustice.

—BACH & BELARDO

The General Theory of Information asserts that reality consists of an infinite amount of inseparable information, as envisioned by John Archibald Wheeler, who argued that "Information is Everything."[1] This contradicts the commonly accepted assumption that elementary particles and matter constitute the foundation of reality.[9] In contrast to the natural sciences' concern with prediction and control,[18] the General Theory of Information is concerned with explaining and understanding that which cannot be readily observed. Inseparability, however, is the essential factor that breaks down the barriers that have separated humanity from nature and human beings from one another. The General Theory of Information is based on the premise that science should fully serve humanity by enabling genuine collaboration and ensuring inclusivity. Inseparability implies that responsibility is indivisible, that we are collectively responsible for the state of the world and that we are individually responsible for making the world a better place. The fruits of our good and positive actions will let us get there together. We have the ability to know it all, because we consist of it all, without the need to believe, and by choice. If we only take responsibility, all is well. This is the hope.

The General Theory of Information also helps to foster mental prosperity by stopping the prevalent thinking that we are only small parts of a big machine that we cannot control and for which we have no responsibility for anyway.

Quite the contrary is true. The General Theory of Information demonstrates that a sustainable life in peace is at our fingertips: a healthy sustainable environment, freedom from personal harm, and happiness in a worldwide inclusive system as each other's equals.

Waiting for Godot: Waiting for Somebody Else to Do Our Jobs

The people have the authority and absolute responsibility. Now they have to live up to it!

—BACH & BELARDO

Why are we always waiting for somebody else to do our jobs? Because we feel insignificant and therefore hope that somebody else will take responsibility for us—helping us, loving us, doing our jobs for us.

The General Theory of Information advises differently: each of us is fully capable of building paradise. Period. No further thinking required! Why? Because when you start thinking about how to build paradise, you will lose hope. Just do it—Godot's last words!

Readings

1. Siegfried, T. The bit and the pendulum: From quantum computing to M theory—The new physics of information. (John Wiley & Sons, New York; 1999).
2. Everett III, H. "Relative state" formulation of quantum mechanics. *Review of Modern Physics* 29, 454–462 (1957).
3. Aspect, A. Bell's inequality test: more ideal than ever. *Nature* 398, 189–190 (1999).
4. Howard, D. in Philosophical Consequences of Quantum Theories: Reflections on Bell's Theorem. (eds. J. Cushing & E. McMullis) 224–253 (University of Notre Dame Press, Notre Dame; 1989).
5. Belousek, D. W. Bell's Theorem, nonseparability, and space-time individuation in quantum mechanics. *Philosophy of Science* 66, S28–47 (1999).
6. Svozil, K. Conventions in Relativity and Quantum Mechanics. *Foundations of Physics* 32, 479–502 (2002).
7. Bohm, D. J. & Hiley, B.J. The undivided universe. (Routledge, London; 1993).
8. Bohm, D. Wholeness and the implicate order. (Routledge, New York; 2002).
9. Healey, R. A. Holism and nonseparability. *The Journal of Philosophy* 88, 393–421 (1991).
10. Kuhn, T. S. The Structure of Scientific Revolutions, Second Edition. (The University of Chicago Press, Chicago; 1970).
11. Bell, J. S. Speakable and unspeakable in quantum mechanics. (Cambridge University Press, Cambridge; 1987).
12. Bell, J., S. Bertlmann's socks and the nature of reality. *Journal de Physique Colloques* 42, C2-41-C42-62 (1981).
13. Bach, C. & Belardo, S. in Proceedings of the Ninth Americas Conference on Information Systems 2629–2639 (Association for Information Systems, August 4–6, Tampa, FL; 2003).

14. Shannon, C.E. A mathematical theory of communication. *The Bell System Technical Journal* 27, 379–423 and 623–656 (1948).
15. Einstein, A., Podolsky, B. & Rosen, N. Can quantum-mechanical description of physical reality be considered complete? *The Physical Review (New York)* 47, 777–780 (1935).
16. Davies, P. C. W. & Brown, J. R. The ghost in the atom: a discussion of the mysteries of quantum physics. (Cambridge University Press, Cambridge; 1986).
17. Bell, J. S. On the Einstein-Podolsky-Rosen paradox. *Physics* 1, 195–200 (1964).
18. Kerlinger, F. N. & Lee, H. B. in Foundation of Behavioral Research, 4th ed., Edn. 4th edition. (eds. F. N. Kerlinger & H. B. Lee) 3–21 (Thomson Learning, New York; 2000).
19. Davies, P. God and the new physics. (Dent & Sons, London; 1986).
20. Renshaw, C. Moving Clocks, Reference Frames and the Twin Paradox. *IEEE Aerospace and Electronic Systems Magazine* 11 (1996).
21. Bohm, D. J. & Hiley, B. J. On the intuitive understanding of nonlocality as implied by quantum theory. *Foundations of Physics* 5, 93–109 (1975).
22. Polanyi, M. Personal knowledge: toward a post-critical philosophy. (Harper Torchbooks, New York; 1962).
23. Kerlinger, F. N. & Lee, H. B. Foundation of Behavioral Research, 4th ed., Edn. 4th edition. (Thomson Learning, New York; 2000).
24. Eigen, M. The origin of genetic information: viruses as models. *Gene* 135, 37–47 (1993).
25. Eigen, M. & Schuster, P. Stages of emerging life—Five principles of early organization. *Journal of Molecular Evolution* 19, 47–61 (1982).

26. Eigen, M., Biebricher, C. K., Gebinoga, M. & Gardiner, W. C. The hypercycle. Coupling of RNA and protein biosynthesis in the infection cycle of an RNA bacteriophage. *Biochemistry* 30, 11005-11018 (1991).
27. Weirauch, M. T. & Hughes, T. R. Conserved expression without conserved regulatory sequence: the more things change, the more they stay the same. *Trends in Genetics* 26, 66–74 (2010).
28. Jagers op Akkerhuis, G. Towards a Hierarchical Definition of Life, the Organism, and Death. *Foundations of Science* 15, 245–262 (2010).
29. Iyer, B., Kenward, M. & Arya, G. Hierarchies in eukaryotic genome organization: Insights from polymer theory and simulations. *BMC Biophysics* 4, 8 (2011).
30. Yin, P., Choi, H. M. T., Calvert, C. R. & Pierce, N. A. Programming biomolecular self-assembly pathways. *Nature* 451, 318–322 (2008).
31. Andersen, E. S. et al. Self-assembly of a nanoscale DNA box with a controllable lid. *Nature* 459, 73–76 (2009).
32. Castro, C. E. et al. A primer to scaffolded DNA origami. *Nature Methods* 8, 221–229 (2011).
33. Rothemund, P. W. K. Folding DNA to create nanoscale shapes and patterns. *Nature* 440, 297–302 (2006).
34. He, Y. et al. Hierarchical self-assembly of DNA into symmetric supramolecular polyhedra. *Nature* 452, 198–201 (2008).
35. Demaine, E. D. & O'Rourke, J. Geometric Folding Algorithms: Linkages, Origami, Polyhedra. (Cambridge University Press Boston; 2007).